Geometric Inequalities

Mathematical Olympiad Series

ISSN: 1793-8570

Published

Vol. 4 Combinatorial Problems in Mathematical Competitions
by Yao Zhang (Hunan Normal University, P. R. China)

Vol. 5 Selected Problems of the Vietnamese Olympiad (1962–2009)
by Le Hai Chau (Ministry of Education and Training, Vietnam)
& Le Hai Khoi (Nanyang Technology University, Singapore)

Vol. 6 Lecture Notes on Mathematical Olympiad Courses:
For Junior Section (In 2 Volumes)
by Xu Jiagu

Vol. 7 A Second Step to Mathematical Olympiad Problems
by Derek Holton (University of Otago, New Zealand &
University of Melbourne, Australia)

Vol. 8 Lecture Notes on Mathematical Olympiad Courses:
For Senior Section (In 2 Volumes)
by Xu Jiagu

Vol. 9 Mathemaitcal Olympiad in China (2009–2010)
edited by Bin Xiong (East China Normal University, China) &
Peng Yee Lee (Nanyang Technological University, Singapore)

Vol. 11 Methods and Techniques for Proving Inequalities
by Yong Su (Stanford University, USA) &
Bin Xiong (East China Normal University, China)

Vol. 12 Geometric Inequalities
by Gangsong Leng (Shanghai University, China)
translated by: Yongming Liu (East China Normal University, China)

The complete list of the published volumes in the series can be found at
http://www.worldscientific.com/series/mos

Gangsong Leng
Shanghai University, China

translated by
Yongming Liu
East China Normal University, China

Vol. 12 Mathematical Olympiad Series

Geometric Inequalities

Published by

East China Normal University Press
3663 North Zhongshan Road
Shanghai 200062
China

and

World Scientific Publishing Co. Pte. Ltd.
5 Toh Tuck Link, Singapore 596224
USA office: 27 Warren Street, Suite 401-402, Hackensack, NJ 07601
UK office: 57 Shelton Street, Covent Garden, London WC2H 9HE

British Library Cataloguing-in-Publication Data
A catalogue record for this book is available from the British Library.

Mathematical Olympiad Series — Vol. 12
GEOMETRIC INEQUALITIES

ISBN 978-981-4704-13-7
ISBN 978-981-4696-48-7 (pbk)

Printed in Singapore.

Contents

Contents

Preface

"God is always doing geometry", said Plato. But the deep investigation and extensive attention to geometric inequalities as an independent field is a matter of modern times.

Many geometric inequalities are not only typical examples of mathematical beauty but also tools for application as well. The well known Brunn-Minkowski's inequality is such an example. "It is like a large octopus, whose tentacles stretches out into almost every field of mathematics. It has not only relation with advanced mathematics such as the Hodge index theorem in algebraic geometry, but also plays an important role in applied subjects such as stereology, statistical mechanics and information theory".

There are dozens of books on geometric inequalities so far, in which "Geometric Inequalities" by Yu. D. Burago and V. A. Zalgaller is cited worldwide. And "Geometric Inequalities" by Chinese scholar Mr. San Zhun is an excellent introductory book (Shanghai Education Press, 1980).

The aim of this book is mainly to introduce geometric inequalities to students and high school teachers who wish to attend the Mathematics Olympiad Competition. The material is elementary. In the process of writing, I strive to achieve: firstly, carefully select new achievements, methods and techniques of recent studies. Secondly, the material should be simple but non-trivial, with an interesting and profound background. Thirdly, as far as possible to present the students' excellent answers and, of course, also some results on my research and experiences. Any comments and suggestions are welcome.

I dedicate this book to Mr. Qiu Zonghu as a form of congratulations on his seventieth birthday and also to commemorate his great contributions to the Mathematical Olympiad of China.

At last, I would like to thank Mr. Ni Ming for his faithfulness and patience in the publication of this book. And thanks also give to my doctoral student Mr. Si Lin for his typist and drawings.

My cherished desire is that the readers like this book.

Leng Gangsong
October 2004

Chapter 1 The method of segment replacement for distance inequalities

The comparison of lengths is more basic than comparison of other geometric quantities (such as angles, areas and volumes). A geometric inequality that involves only the lengths is called a distance inequality.

Some simple axioms and theorems on inequalities in Euclidean geometry are usually the starting point to solve problems of distance inequality, in which most frequently used tools are:

Proposition 1. The shortest line connecting point A with point B is the segment AB.

The direct corollary of Proposition 1 is

Proposition 2 (Triangle Inequality). For arbitrary three points A, B and C, we have $AB \leqslant AC + CB$, the equality holds if and only if C is on the segment AB.

Remark. In this book, to simplify notations, any symbol of geometric object also denotes its quantity according to the context.

Proposition 2 has the following often used consequences.

Proposition 3. In a triangle, the longer side has the larger opposite angle. And the larger angle has the longer opposite side.

Proposition 4. The median of a triangle on a side is shorter than the half-sum of the other two sides.

Proposition 5. If a convex polygon is within another one, then the outside convex polygon's perimeter is larger.

Proposition 6. Any segment in a convex polygon is either less than the longest side or the longest diagonal of the convex polygon.

Firstly, we give an example.

Example 1. Let a, b and c be sides of $\triangle ABC$. Prove that

$$\frac{a}{b+c}+\frac{b}{c+a}+\frac{c}{a+b}<2. \tag{1}$$

Proof. By the triangle inequality $a<b+c$, yields

$$\frac{a}{b+c}=\frac{2a}{2(b+c)}<\frac{2a}{a+b+c}.$$

Similarly,

$$\frac{b}{c+a}<\frac{2b}{a+b+c},$$

$$\frac{c}{b+a}<\frac{2c}{a+b+c}.$$

Adding up the above three inequalities leads to Inequality (1). □

Example 2. Let AB be the longest side of $\triangle ABC$, and P a point in the triangle, prove that

$$PA+PB>PC. \tag{2}$$

Proof. Let D be the intersection point of CP and AB (see Figure 1.1), then $\angle ADC$ or $\angle BDC$ is not acute. Without loss of generality, we assume that $\angle ADC$ is not acute. Applying Proposition 3 to $\triangle ADC$, we obtain

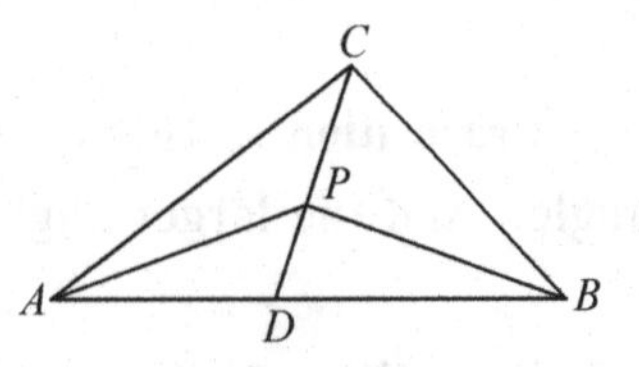

Figure 1.1

$$AC \geqslant CD.$$

Therefore,

$$AB \geqslant AC \geqslant CD > PC. \tag{a}$$

Furthermore, applying triangle inequality to $\triangle PAB$, we have

$$PA + PB > AB. \tag{b}$$

Combining (a) and (b), we obtain Inequality (2) immediately. □

Remark. (1) If AB is not the longest, then the conclusion may not be true.

(2) If point P on the plane of regular $\triangle ABC$, and P is not on the circumcircle of the triangle, then the sum of any two of PA, PB and PC is longer than the remaining one. That is, PA, PB and PC consist of a triangle's three sides.

Example 3. A closed polygonal line with perimeter 1 can be put inside a circle with radius $\frac{1}{4}$.

Analysis. The key to prove Example 3 is to determine the center O of the circle, such that the distance of point on the polygonal line to point O is less or equal to $\frac{1}{4}$.

Proof. Let points A and B be two arbitrary points that bisect the perimeter of the closed polygonal line (see Figure 1.2). That is, the length of the polygonal line $\widehat{AB}$ is $\frac{1}{2}$. Let the center of circle O be the midpoint of the segment AB, then the distance from each point on the closed polygonal line to point O is less than $\frac{1}{4}$.

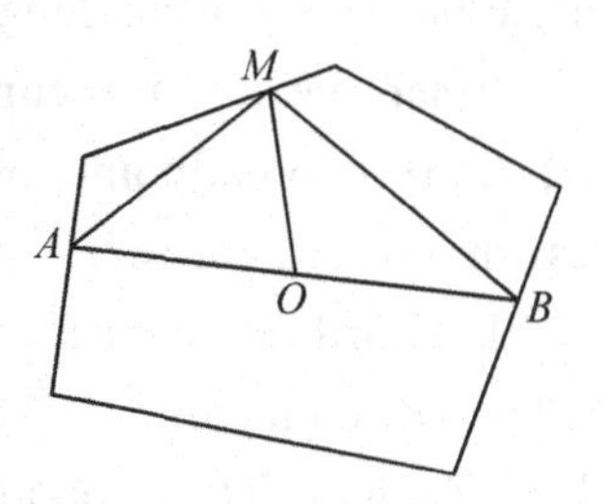

Figure 1.2

In fact, let M be a point on the closed polygonal line but not A or

B, applying Proposition 4, we have

$$OM < \frac{1}{2}(AM + MB) \leqslant \frac{1}{2}(\widehat{AM} + \widehat{BM}) < \frac{1}{2}\widehat{AB} = \frac{1}{4},$$

where symbols such as $\widehat{AM}$ denote the polygonal line with endpoints A and B, and its length as well.

And if M is A or B, then $OM = \frac{AB}{2} < \frac{1}{4}$ by Proposition 1.

Now, we draw a circle with center O and radius $\frac{1}{4}$, then the polygon is located inside the circle. □

In fact, the proofs of above examples embody an idea of "segment replacement", we call it "the method of segment replacement". Specifically, this method is based on Proposition 1 or its inference, replace curve to polygonal line, then replace polygonal line to segment. This method is one of the most commonly used methods for proving geometric inequalities, especially distance inequalities.

Now, here are other examples.

Firstly, we introduce the classical Pólya's problem.

Example 4. Of all the curves that can bisect the area of a circle and with their endpoints on its circumference, the diameter of the circle has the shortest length.

Proof. Denote the curve by $\overset{\frown}{AB}$, points A, B on the circle. If A and B are two endpoints of a diameter, then it is clear that $\overset{\frown}{AB}$ is not less than the diameter.

If chord AB is not a diameter (see Figure 1.3), let diameter CD be parallel to AB (if $A = B$, then CD can be any diameter that does not intersect with A or B), then curve $\overset{\frown}{AB}$ intersects CD at point E which is not the center, hence

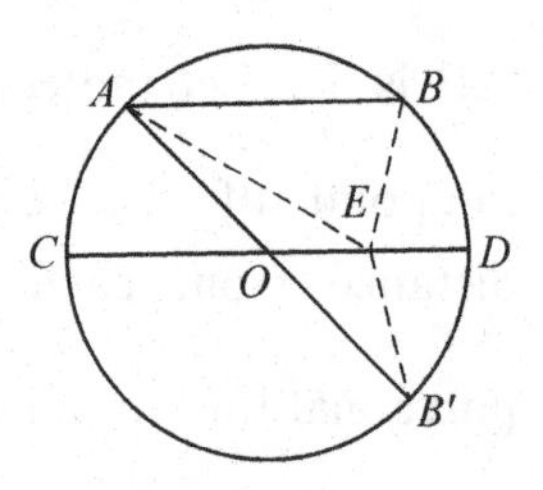

Figure 1.3

$$\overset{\frown}{AB} = \overset{\frown}{AE} + \overset{\frown}{EB} \geqslant AE + EB.$$

The key point is that we consider the polygonal line instead of the curve.

Now we show that $\overset{\frown}{AE} + \overset{\frown}{EB} >$ the diameter of the circle. Let B' be the symmetric point of B to CD, then AB' is the diameter of the circle. So

$$AE + EB = AE + EB' > AB' = \text{the diameter of the circle.}$$

□

The following example stems from our research on extremal property of the pedal triangle.

Example 5. Let P be a point in $\triangle ABC$, and A', B' and C' be the projection of P onto BC, CA and AB or their extended, respectively; Let A'', B'' and C'' be the intersection points of AP, BP and CP to corresponding sides, respectively. And the perimeter of $\triangle A''B''C''$ equals 1. Prove that

$$\widehat{A'B''C'A''} + \widehat{A'C''B'A''} \leqslant 2.$$

Proof. The required inequality is equivalent to

$$A'B'' + B''C' + C'A'' + A'C'' + C''B' + B'A'' \leqslant 2. \tag{a}$$

To prove (a), we need only to prove the local inequality

$$A'B'' + A'C'' \leqslant A''B'' + A''C''. \tag{b}$$

Similarly,

$$B'A'' + B'C'' \leqslant B''A'' + B''C'',$$
$$C'A'' + C'B'' \leqslant C''A'' + C''B''.$$

Adding up these inequalities, we get (a) immediately. □

Before proceeding to prove (b) we need the following lemma.

Lemma 1. Let point P be on the altitude AD of $\triangle ABC$ (see Figure

1.4). If BP intersects AC at E, and CP intersects AB at F, then

$$\angle FDA = \angle EDA.$$

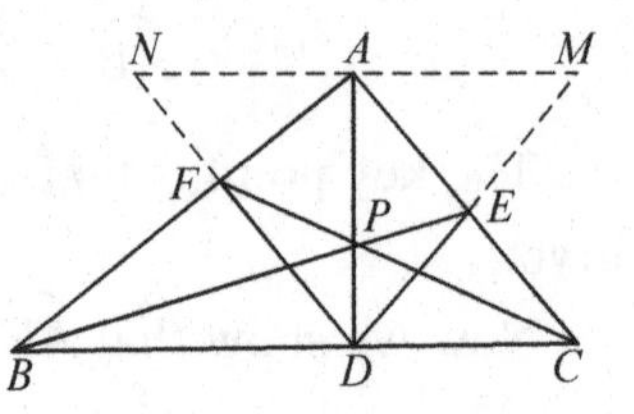

Figure 1.4

Proof. Suppose that the line parallel to BC intersects lines DE and DF at points M and N, respectively, then

$$\frac{AF}{BF} = \frac{AN}{BD}, \frac{CE}{AE} = \frac{CD}{AM}.$$

By Ceva's theorem,

$$\frac{AF}{BF} \cdot \frac{BD}{DC} \cdot \frac{CE}{EA} = 1,$$

that is,

$$AM = AN.$$

By $AD \perp MN$, we have $DM = DN$, so

$$\angle EDA = \angle ADM = \angle ADN = \angle FDA.$$

□

Now we turn to prove (b):

Proof. (i) If P is a point on the altitude AD of $\triangle ABC$, then $A' = A''$, obviously, (b) is true.

(ii) If P is not a point on the altitude AD of $\triangle ABC$ (see Figure 1.5), without loss of generality, we may assume that P, B lie on ipsilateral of AB, line $A'P$ intersects AB at M, MC intersects BB'' at M''. By Lemma 1 we have

$$\angle B''A'P > \angle M'A'P = \angle C''A'P. \quad (c)$$

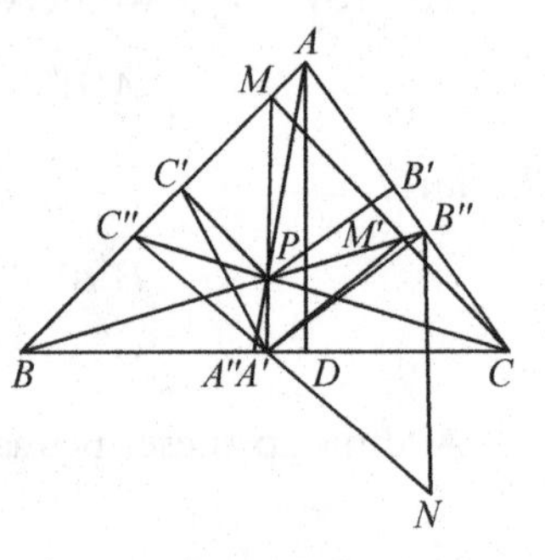

Figure 1.5

Let N be the symmetric point of B'' to BC, then

$$\angle NA'C = \angle CA'B''.$$

By (c) we have

$$\begin{aligned} &\angle NA'C + \angle C''A'C \\ =\ &\angle NA'C + \angle C''A'P + \angle PA'C \\ <\ &\angle PA'B'' + \angle PA'N \\ =\ &\pi, \end{aligned}$$

so A' and A'' lie on ipsilateral of $C''N$, that is, A' is a point in $\triangle C''A''N$, then by Proposition 5 we have

$$A''C'' + A''N > A'C'' + A'N.$$

Notice that $A''B'' = A''N$, $A'B'' = A'N$, we have

$$A''B'' + A''C'' \geqslant A'B'' + A'C''.$$

Thus, we have proved (b). □

Remark. (1) Symmetric reflection method in Example 5 is an often used means of segment replacement.

(2) Applying inequality (b), Dr. Yuan Jun proved a conjecture of Mr. Liu Jian:

The perimeter of $\triangle A'B'C' \leqslant$ the perimeter of $\triangle A''B''C''$.

The following example is a rather hard problem.

Example 6. Let P be a point in $\triangle ABC$, show that

$$\sqrt{PA} + \sqrt{PB} + \sqrt{PC} < \frac{\sqrt{5}}{2}(\sqrt{BC} + \sqrt{CA} + \sqrt{AB}). \qquad \text{(a)}$$

First we state the following lemma which can be derived by Proposition 5 directly.

Lemma 2. Let P be a point in the convex quadrilateral $ABCD$, then

$$PB + PC < BA + AD + DC.$$

Next we prove (a).

Proof. Let $BC = a$, $AC = b$, $BA = c$, $PA = x$, $PB = y$ and $PC = z$ (see Figure 1.6), and let A', B' and C' be midpoints of the sides of $\triangle ABC$, then P must be in one of the parallelograms $A'B'AC'$, $C'B'CA'$ and $B'A'BC'$. Without loss of generality, we can assume that P is in parallelogram $A'B'AC'$, then applying Lemma 2 to convex quadrilateral $ABA'B'$, we have

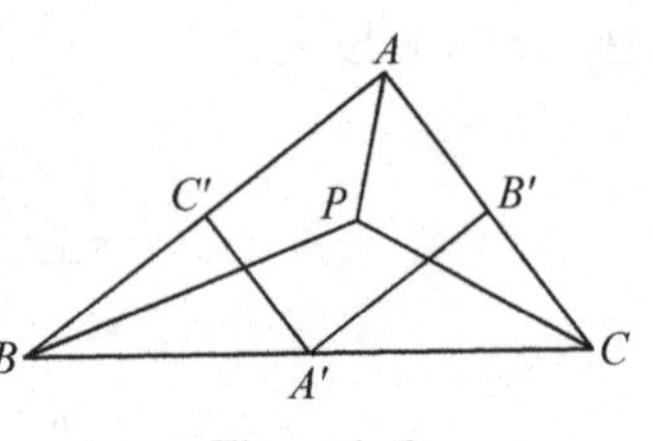

Figure 1.6

$$PA + PB < BA' + A'B' + B'A,$$

that is

$$x + y < \frac{1}{2}(a + b + c). \tag{b}$$

Similarly, for convex quadrilateral $ACA'C'$, we have

$$PA + PC < AC' + C'A' + A'C,$$

that is

$$x + z < \frac{1}{2}(a + b + c). \tag{c}$$

Adding up (b) and (c), we find that

$$2x + y + z < a + b + c. \tag{d}$$

Now we notice that the original inequality is equivalent to

$$(\sqrt{x} + \sqrt{y} + \sqrt{z})^2 < \frac{5}{4}(\sqrt{a} + \sqrt{b} + \sqrt{c})^2,$$

that is

$$\begin{aligned} & x + y + z + 2\sqrt{xy} + 2\sqrt{xz} + 2\sqrt{yz} \\ & < \frac{5}{4}(a + b + c + 2\sqrt{ab} + 2\sqrt{bc} + 2\sqrt{ac}). \end{aligned} \tag{e}$$

Thus, it suffices to prove (e).

By the mean value inequality, we have

$$2\sqrt{xy} \leqslant 2x + \frac{1}{2}y,$$

$$2\sqrt{xz} \leqslant 2x + \frac{1}{2}z,$$

$$2\sqrt{yz} \leqslant y + z.$$

Combining these three inequalities and inequality (d), we get

$$\text{the left side of (e)} \leqslant x + y + z + 2x + \frac{1}{2}y + 2x + \frac{1}{2}z + y + z$$
$$= \frac{5}{2}(2x + y + z) < \frac{5}{2}(a + b + c),$$

so we need only to prove

$$\frac{5}{2}(a + b + c) < \frac{5}{4}(a + b + c + 2\sqrt{ab} + 2\sqrt{bc} + 2\sqrt{ac}). \quad \text{(f)}$$

But (f) is equivalent to

$$a + b + c < 2(\sqrt{ab} + \sqrt{bc} + \sqrt{ac}), \quad \text{(g)}$$

which is a simple inequality. In fact, without loss of generality, suppose that $a \geqslant b \geqslant c$, then by $b + c > a$,

the right side of (g) $> 2(b + c) > a + b + c =$ the left side of (g).

Therefore, (a) has been proved. □

Remark. (1) The constant $\sqrt{5}/2$ of inequality (a) is optimal, the proof of which is left to the reader.

(2) The elegant answer above was given by Zhu Qingsan (the former student of High School Affiliated to South China Normal University, who won a gold medal at the 45th IMO in 2004). Smartly positioning point P and dealing well with the non-fully symmetry variable are the key points to the answer.

Of course, Example 6 can also be proved by contour line. Contour line is a special plane curve, such as circle, ellipse and so on, introduced to discuss extremal problems. Here we use ellipse as the contour line.

An Alternative Proof of Example 6

Proof. Let $BC = a$, $CA = b$ and $AB = c$. Without loss of generality, suppose that $a \leqslant b$ and $a \leqslant c$.

Now, we make an ellipse through P with focal points B and C, and intersects AB and AC at E and F, respectively (see Figure 1.7), then by Proposition 1 we have

$$PA \leqslant \max(EA, FA).$$

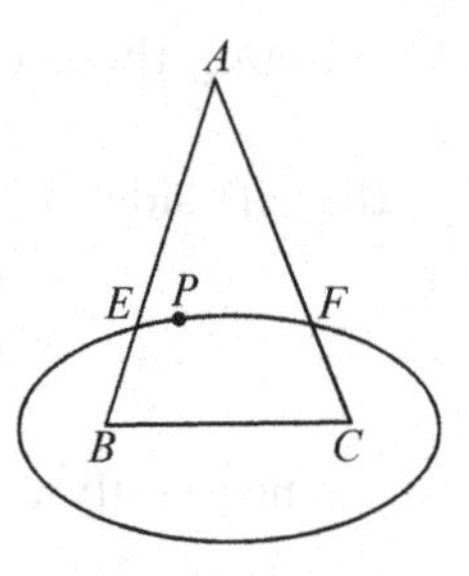

Figure 1.7

Without loss of generality, suppose that $EA \geqslant FA$, then $PA \leqslant EA$. Further,

$$\sqrt{PC} + \sqrt{PB} \leqslant \sqrt{2(PB + PC)} = \sqrt{2(EB + EC)},$$

therefore

$$\begin{aligned}
&\sqrt{PA} + \sqrt{PB} + \sqrt{PC} \\
&< \sqrt{EA} + \sqrt{2(EB + EC)} \\
&\leqslant \left[5EA + \frac{5}{2}(EB + EC)\right]^{\frac{1}{2}} \\
&= \left[5(EA + EB) + \frac{5}{2}(EC - EB)\right]^{\frac{1}{2}} \\
&< \sqrt{5}\sqrt{c + \frac{a}{2}} \\
&< \frac{\sqrt{5}}{2}(\sqrt{a} + \sqrt{b} + \sqrt{c}).
\end{aligned}$$

□

Exercises 1

1. Suppose that A' is a point on bisector AT of the exterior angle of $\triangle ABC$, show that

$$A'B + A'C \geqslant AB + AC.$$

2. Given sides $a > b > c$ of $\triangle ABC$ and any point O in $\triangle ABC$, suppose that the lines AO, BO and CO intersect a, b and c at P, Q and R, respectively. Prove that

$$OP + OQ + OR < a.$$

3. Let D, E and F be points on sides BC, CA and AB of $\triangle ABC$, respectively. Let Δ and R be the area and circumcircle radius of $\triangle ABC$, respectively. Show that

$$DE + EF + FD \geqslant \frac{2\Delta}{R}.$$

4. Let E and F be two points on rays AC and AB. Show that

$$|AB - AC| + |AE - AF| \geqslant |BE - CF|,$$

the equality holds if and only if $AB = AC$ and $AE = AF$.

5. Let O be a point in hexagon $A_1A_2A_3A_4A_5A_6$, such that $\angle A_iOA_{i+1} = \frac{\pi}{3}$, $(i = 1, 2, 3, 4, 5, 6)$ $(A_7 = A_1)$. If $OA_1 > OA_3 > OA_5$, $OA_2 > OA_4 > OA_6$, then

$$A_1A_2 + A_3A_4 + A_5A_6 < A_2A_3 + A_4A_5 + A_6A_1.$$

Chapter 2 Ptolemy's inequality and its application

The famous Ptolemy's inequality is a distance inequality for arbitrary quadrilateral. It can be written as

Theorem 1 (Ptolemy's inequality). In the quadrilateral $ABCD$, we have

$$AB \cdot CD + AD \cdot BC \geqslant AC \cdot BD,$$

the equality holds if and only if four points A, B, C and D lie on a circle.

Proof. Let E be a point in quadrilateral $ABCD$ (see Figure 2.1), such that $\angle BAE = \angle CAD$, $\angle ABE = \angle ACD$, then $\triangle ABE \backsim \triangle ACD$. So $AB \cdot CD = AC \cdot BE$. Also $\angle BAC = \angle EAD$, and $AB/AE = AC/AD$, then $\triangle ABC \backsim \triangle AED$, $AD \cdot BC = AC \cdot DE$. So

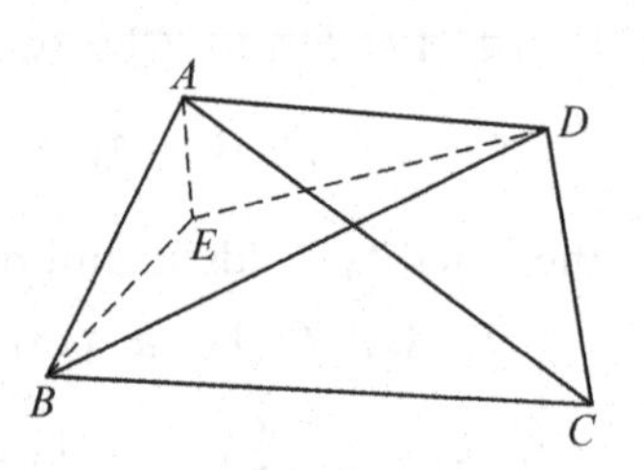

Figure 2.1

$$AB \cdot CD + AD \cdot BC = AC(BE + DE) \geqslant AC \cdot BD,$$

the equality holds if and only if point E is on segment BD. Thus $\angle ABD = \angle ACD$, so $ABCD$ is an inscribed quadrilateral. □

By applying Ptolemy's inequality, we have simple proofs for some distance inequalities.

Example 1 (Klamkin's dual inequality). Let a, b and c be the three sides of $\triangle ABC$, and let m_b and m_c be medians of b and c, respectively.

Prove that

$$4m_a m_b \leqslant 2a^2 + bc. \tag{1}$$

Proof. Construct parallelogram $ABCD$ and $ACBE$ (see Figure 2.2), connect BD and CE. Notice that $DE = 2a$, $BD = 2m_a$, and $CE = 2m_c$, applying Ptolemy's inequality for quadrilateral $BCDE$,

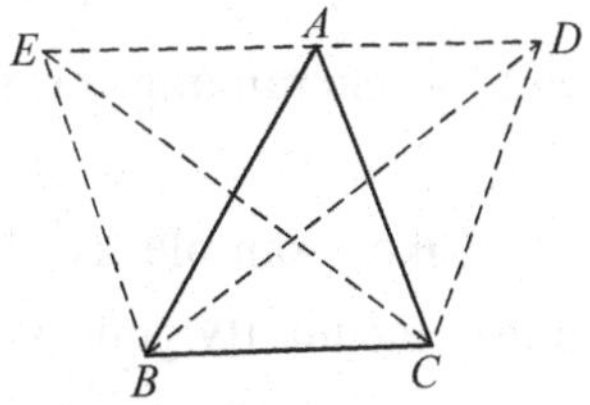

$$BC \cdot DE + BE \cdot CD \geqslant BD \cdot EC.$$

Figure 2.2

That is (1). □

Example 2. Let a, b and c be the three sides of $\triangle ABC$, and let m_a, m_b, and m_c be medians on sides a, b and c, respectively. Prove that

$$m_a(bc - a^2) + m_b(ac - b^2) + m_c(ab - c^2) \geqslant 0. \tag{2}$$

The key to the following proof is to find a special quadrilateral.

Proof. Let AD, BE and CF be medians of triangle ABC with barycenter G (see Figure 2.3).

Applying Ptolemy's inequality to quadrilateral $BDGF$,

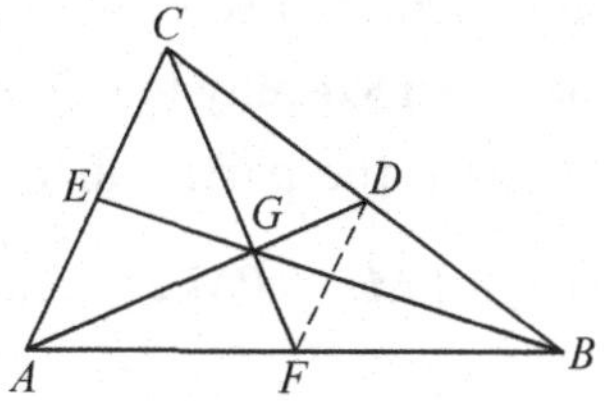

$$BG \cdot DF \leqslant GF \cdot DB + DG \cdot BF. \text{ (a)}$$

Figure 2.3

Notice that $BG = \frac{2}{3}m_b$, $DG = \frac{1}{3}m_a$, $GF = \frac{1}{3}m_c$ and $DF = \frac{b}{2}$, (a) can be rewritten into:

$$2bm_b \leqslant am_c + cm_a.$$

So

$$2b^2 m_b \leqslant abm_c + cbm_a. \tag{b}$$

Similarly,

$$2c^2 m_c \leqslant acm_b + bcm_a, \tag{c}$$

$$2a^2 m_a \leqslant abm_c + acm_b. \tag{d}$$

Adding up (b), (c) and (d), we have

$$2(m_a bc + m_b ca + m_c ab) \geqslant 2(m_a a^2 + m_b b^2 + m_c c^2),$$

and by rearranging terms, we obtain inequality (2). □

Like Example 2, the following is another example of geometric linear inequality generated by Ptolemy's theorem.

Example 3. Let $A_1A_2\cdots A_n$ be a regular n-polygon, and M_1, M_2, ..., M_n be midpoints of the corresponding sides. Let P be an arbitrary point in the plane which n-polygon lies in.

Prove that

$$\sum_{i=1}^{n} PM_i \geqslant \left(\cos\frac{\pi}{n}\right)\sum_{i=1}^{n} PA_i. \tag{3}$$

Proof. Let M_{i-1}, M_i be midpoints of the $(i-1)$th and (i)th edge of the regular n-polygon, respectively (see Figure 2.4). Applying Ptolemy's inequality to quadrilateral $PM_{i-1}A_iM_i$, we obtain the partial inequality

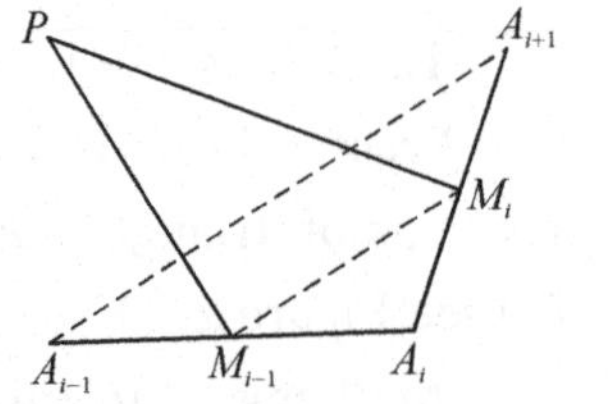

Figure 2.4

$$A_iM_{i-1} \cdot PM_i + PM_{i-1} \cdot A_iM_i \geqslant PA_i \cdot M_{i-1}M_i,$$

so

$$PM_i + PM_{i-1} \geqslant 2\left(\cos\frac{\pi}{n}\right) \cdot PA_i, \tag{a}$$

where $i = 1, 2, \ldots, n$ and $A_0 = A_n$, $M_0 = M_n$.

Now summation on both sides of (a), we have

$$\sum_{i=1}^{n}(PM_i + PM_{i-1}) \geqslant 2\left(\cos\frac{\pi}{n}\right) \cdot \sum_{i=1}^{n} PA_i,$$

which is equivalent to inequality (3). □

The following two examples introduce skills of dealing with

inequalities involving moving point in or out of quadrilateral or triangle by Ptolemy's theorem.

Example 4. Let P be a point in parallelogram $ABCD$, prove that

$$PA \cdot PC + PB \cdot PD \geqslant AB \cdot BC, \tag{4}$$

and point out the condition that the equality holds.

Proof. Let PQ be parallel and equal to CD, connect CQ, BQ (see Figure 2.5), then $CDPQ$ and $ABQP$ are parallelograms. So

$$CQ = PD,\ BQ = PA,\ PQ = AB.$$

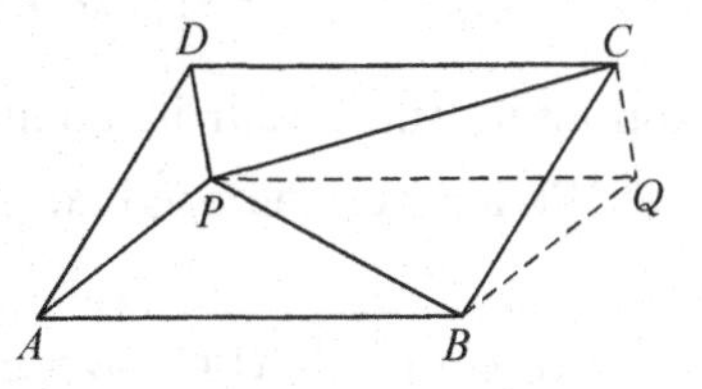

Figure 2.5

Applying Ptolemy's inequality to quadrilateral $PBQC$:

$$BQ \cdot PC + PB \cdot CQ \geqslant PQ \cdot BC,$$

that is

$$PA \cdot PC + PB \cdot PD \geqslant AB \cdot BC,$$

equality holds if and only if P, B, Q, C are on a circle, that is $\angle CPB + \angle CQB = \pi$ and $\angle CQB = \angle APD$, so equality of (4) holds for

$$\angle APD + \angle CPB = \pi.$$

□

Example 5. In $\triangle ABC$, $\angle A = 60°$, let P be a point in the plan which $\triangle ABC$ lie in, such that $PA = 6$, $PB = 7$ and $PC = 10$. Find the maximal area of $\triangle ABC$.

Answer 1: We first prove a lemma.

Lemma 1. In the convex quadrilateral $XYZU$, the diagonal XZ and YU intersect at point O, and let $\angle XOY = \theta$, then

$$YZ^2 + UX^2 - XY^2 - ZU^2 = 2XZ \cdot YU \cdot \cos\theta.$$

Proof. Applying the cosine law to $\triangle OYZ$, $\triangle OUX$, $\triangle OXY$ and $\triangle OZU$, respectively (see Figure 2.6). We have

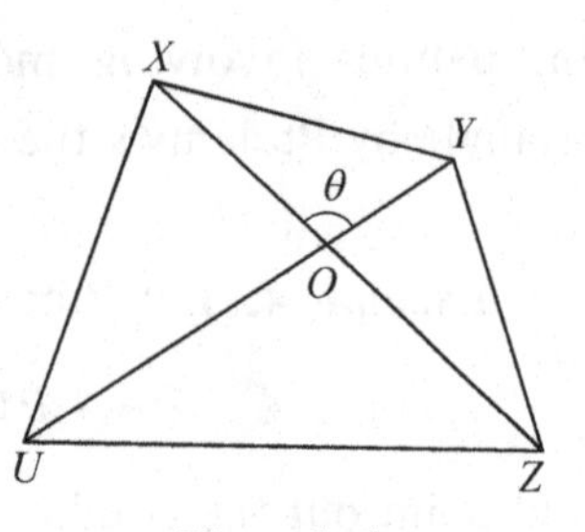

Figure 2.6

$$YZ^2 = OY^2 + OZ^2 + 2OY \cdot OZ \cdot \cos\theta,$$
$$UX^2 = OU^2 + OX^2 + 2OU \cdot OX \cdot \cos\theta,$$
$$XY^2 = OX^2 + OY^2 - 2OX \cdot OY \cdot \cos\theta,$$
$$ZU^2 = OZ^2 + OU^2 - 2OZ \cdot OU \cdot \cos\theta.$$

Adding up these four inequalities, we get Lemma 1 immediately. □

We answer the original problem next.

Proof. In $\triangle ABC$, we construct line through P parallel to AB, and line through A parallel to PB (see Figure 2.7). Let point D be the intersect point of the two parallel lines. Suppose that PD intersect AC at point E, then $\angle CEP = 60^\circ$.

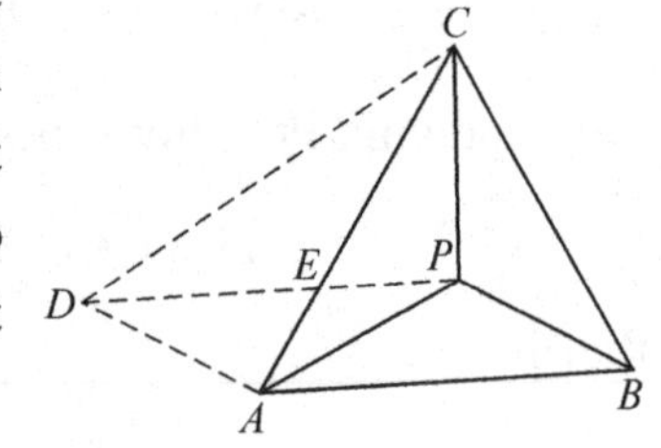

Figure 2.7

Let $AC = x$, $AB = PD = y$ and $CD = t$. Applying Lemma 1 to quadrilateral $APCD$:

$$t^2 + 6^2 - 10^2 - 7^2 = 2\cos 60^\circ \cdot xy,$$

that is

$$xy = t^2 - 113. \tag{a}$$

On the other hand, applying Ptolemy's inequality to quadrilateral $APCD$:

$$xy \leqslant 6t + 70. \tag{b}$$

Combining (1) and (2):

$$t^2 - 6t - 183 \leqslant 0,$$

so

$$0 \leqslant t \leqslant 3 + 8\sqrt{3}. \tag{c}$$

By (c) and (b), we have $xy \leqslant 88 + 48\sqrt{3}$, so

$$S_{\triangle ABC} = \frac{\sqrt{3}}{4} xy \leqslant 36 + 22\sqrt{3},$$

the equality holds if and only if D, A, P and C are on a circle, that is $\angle PBA = \angle PCA$. So the maximal area of $\triangle ABC$ is $36 + 22\sqrt{3}$. □

Answer 2. We first prove a lemma.

Lemma 2. Let P be a point in plane of parallelogram $ABCD$, then

$$PA^2 + PC^2 - PB^2 - PD^2 = 2\overrightarrow{AB} \cdot \overrightarrow{AD}.$$

Proof. Transform $\triangle BPC$ to $\triangle ADP'$ (See Figure 2.8). Let $\overrightarrow{AP} = \alpha$, $\overrightarrow{BP} = \beta$, $\overrightarrow{DP'} = \gamma$, then

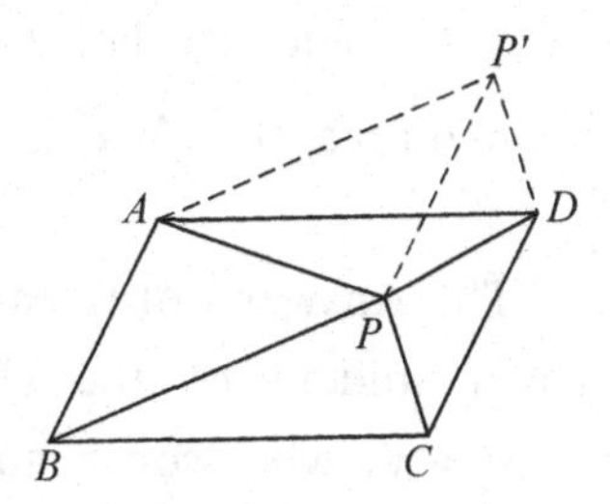

Figure 2.8

$$PA^2 + PC^2 = PA^2 + P'D^2 = \alpha^2 + \gamma^2,$$

$$\begin{aligned} PD^2 + PB^2 &= PD^2 + P'A^2 \\ &= \beta^2 + (\alpha + \beta + \gamma)^2 \\ &= 2\beta^2 + \alpha^2 + \gamma^2 + 2\alpha \cdot \beta \\ &\quad + 2\alpha \cdot \gamma + 2\gamma \cdot \beta, \end{aligned}$$

so

$$\begin{aligned} PA^2 + PC^2 - PD^2 - PB^2 &= -2\beta^2 - 2\alpha \cdot \beta - 2\alpha \cdot \gamma - 2\gamma \cdot \beta \\ &= -2(\alpha + \beta) \cdot (\beta + \gamma) \\ &= 2\overrightarrow{AD} \cdot \overrightarrow{P'P} \\ &= 2\overrightarrow{AB} \cdot \overrightarrow{AD}. \end{aligned}$$

□

We answer the original problem next.

Proof. Translate $\triangle APB$ to $\triangle CP'D$ (see Figure 2.9), then $P'C = 6$, $P'D = 7$, $CD = AB$ and $PP' = AC$.

Let $PD = d$, applying Ptolemy's inequality to quadrilateral $CP'DP$:

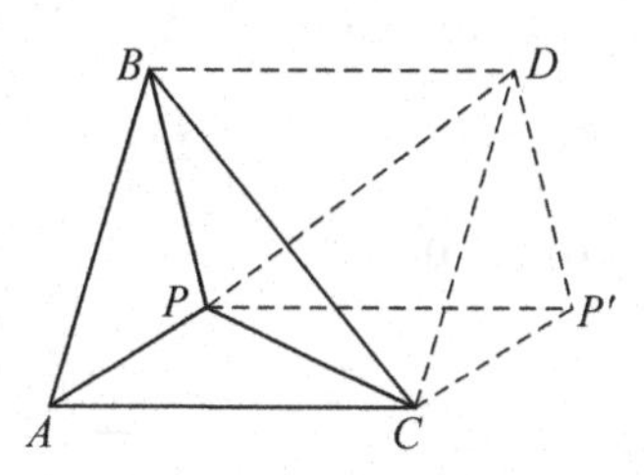

Figure 2.9

$$70 + 6d \geqslant AB \cdot AC. \tag{a}$$

Applying Lemma 2 to parallelogram $ABDC$:

$$7^2 + 10^2 - 6^2 - d^2 = 2\,\overrightarrow{BA} \cdot \overrightarrow{BD} = -AB \cdot AC. \tag{b}$$

Combining (a) and (b):

$$d^2 - 113 \leqslant 6d + 70,$$

so

$$0 \leqslant d \leqslant 3 + 8\sqrt{3}.$$

Thus, by (a), we have $AB \cdot AC \leqslant 88 + 48\sqrt{3}$, then $S_{\triangle ABC} \leqslant 36 + 22\sqrt{3}$, the equality holds if and only if $\angle PBA = \angle PCA$. So the maximal area of $\triangle ABC$ is $36 + 22\sqrt{3}$. □

The Answer 1 of the above example is given by Zhu Xianying (the former student of the High School Affiliated to Hunan Normal University, who won a gold medal at the 45th IMO in 2004), and Answer 2 is given by Zhu Qingsan. The two methods have similarities on the configuration, they are both pretty answers.

The following inequality was given by the famous geometry scientist Bottema. In fact, Ptolemy's inequality holds for quadrilateral in space, so we discuss Bottema's inequality for arbitrary points in space.

Example 6 (Bottema's inequality). Let a_1, a_2, a_3, and b_1, b_2, b_3 be sides of $\triangle A_1A_2A_3$ and $\triangle B_1B_2B_3$ with areas F and F' respectively, and x_1, x_2 and x_3 be the distance from arbitrary point P to A_1, A_2 and A_3, respectively. Let

$$M = b_1^2(-a_1^2 + a_2^2 + a_3^2) + b_2^2(a_1^2 - a_2^2 + a_3^2) + b_3^2(a_1^2 + a_2^2 - a_3^2).$$

Prove that

$$\sum_{i=1}^{3} b_i x_i \geqslant \left(\frac{M}{2} + 8FF'\right)^{\frac{1}{2}}. \tag{5}$$

Proof. Let $A_1A_2 = a_3$, construct $\triangle A_1A_2C$ in the other side of $\triangle A_1A_2A_3$ (see Figure 2.10), such that $\triangle A_1A_2C \backsim \triangle B_1B_2B_3$, then

$$A_1C = \frac{a_3b_2}{b_3},\ A_2C = \frac{a_3b_1}{b_3}.$$

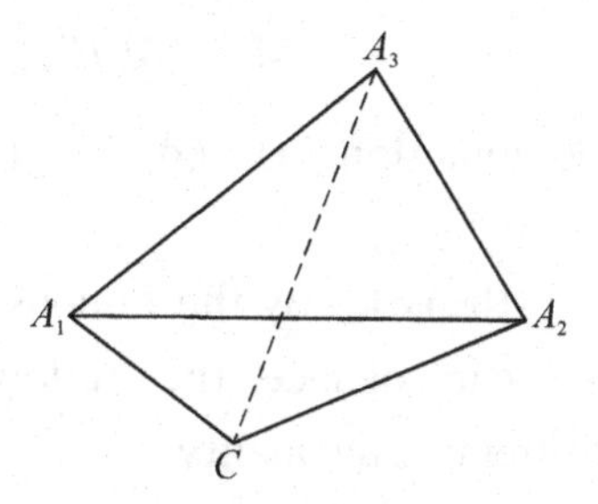

Figure 2.10

Applying Ptolemy's theorem to space quadrilateral PA_1CA_2,

$$x_1\frac{a_3b_1}{b_3} + x_2\frac{a_3b_2}{b_3} \geqslant a_3 \cdot PC,$$

that is

$$b_1x_1 + b_2x_2 \geqslant b_3 \cdot PC.$$

So

$$\begin{aligned} b_1x_1 + b_2x_2 + b_3x_3 &\geqslant b_3 \cdot PC + b_3x_3 \\ &= b_3(PC + x_3) \geqslant b_3 \cdot A_3C, \end{aligned}$$

that is

$$2(b_1x_1 + b_2x_2 + b_3x_3)^2 \geqslant 2b_3^2 \cdot A_3C^2. \tag{a}$$

On the other hand, applying cosine law to $\triangle A_1A_2C$,

$$A_3C^2 = a_2^2 + \left(\frac{a_3b_2}{b_3}\right)^2 - 2a_2 \cdot \frac{a_3b_2}{b_3} \cdot \cos(\angle A_3A_1A_2 + \angle A_2A_1C).$$

So

$$\begin{aligned} 2b_3^2 \cdot A_3C^2 &= 2a_2^2b_3^2 + 2a_3^2b_2^2 - 4a_2a_3b_2b_3\cos(\angle A_3A_1A_2 + \angle A_2A_1C) \\ &= 2a_2^2b_3^2 + 2a_3^2b_2^2 - 4a_2a_3b_2b_3 \cdot (\cos\angle A_3A_1A_2 \cdot \cos\angle A_2A_1C \\ &\quad - \sin\angle A_3A_1A_2 \cdot \sin\angle A_2A_1C) \\ &= 2a_2^2b_3^2 + 2a_3^2b_2^2 - 4a_2a_3b_2b_3 \cdot \frac{a_2^2 + a_3^2 - a_1^2}{2a_2a_3} \cdot \frac{b_2^2 + b_3^2 - b_1^2}{2b_2b_3} \\ &\quad + 4(a_2a_3\sin\angle A_3A_1A_2)(b_2b_3\sin\angle A_2A_1C) \\ &= 2a_2^2b_3^2 + 2a_3^2b_2^2 - (a_2^2 + a_3^2 - a_1^2)(b_2^2 + b_3^2 - b_1^2) + 16FF' \\ &= b_1^2(-a_1^2 + a_2^2 + a_3^2) + b_2^2(a_1^2 - a_2^2 + a_3^2) + b_3^2(a_1^2 + a_2^2 - a_3^2) \\ &\quad + 16FF' \end{aligned}$$

$$= M + 16FF'. \qquad \text{(b)}$$

Combination (a) and (b), Inequality (1) is proved. □

Remark. By the famous Neuberg-Pedoe's inequality: $M \geqslant 16FF'$, we can deduce the following inequality about two triangles by Bottema's inequality

$$b_1x_1 + b_2x_2 + b_3x_3 \geqslant 4\sqrt{FF'}.$$

If $\triangle B_1B_2B_3$ is equilateral, then we can obtain a Fermat's inequality about the distance from interior point to vertex of triangle

$$x_1 + x_2 + x_3 \geqslant 2\sqrt{\sqrt{3}F}.$$

Of course, if $\triangle B_1B_2B_3$ is equilateral, then by Bottema's inequality, a strengthened form of Fermat's inequality can be obtained:

$$x_1 + x_2 + x_3 \geqslant \left(\frac{1}{2}(a^2 + b^2 + c^2) + 2\sqrt{3}F\right)^{\frac{1}{2}}.$$

Exercises 2

1. Let $ABCDEF$ be a convex hexagon, and $AB = BC$, $CD = DE$, $EF = FA$, prove that

$$\frac{BC}{BE} + \frac{DE}{DA} + \frac{FA}{FC} \geqslant \frac{3}{2},$$

and point out the condition that the equality holds.

2. In $\triangle ABC$, let $BC = a$, $AC = b$, and the length of AB be variable, construct a square by AB outside of $\triangle ABC$. Let O be the center of the square, and let M and N be midpoints of BC and AC, respectively. Find the maximum of $OM + ON$.

3. Let $ABCDEF$ be a convex hexagon, and $AB = BC = CD$, $DE = EF = FA$, $\angle BCD = \angle EFA = 60°$. Let G, H be two points in hexagon, such that $\angle AGB = \angle DHE = 120°$. Prove that

$$AG + GB + GH + DH + HE \geqslant CF.$$

(36th IMO problem)

4. Let $\triangle ABC$ be inscribed on $\odot O$, and P be an arbitrary point in $\triangle ABC$. Construct parallel lines of AB, AC, BC through P, intersect BC, AC at F, E, intersect AB, BC at K, I, intersect AB, AC at G, H respectively. Let AD be a chord of $\odot O$ through P, prove that

$$EF^2 + KI^2 + GH^2 \geqslant 4PA \cdot PD.$$

5. In $\triangle ABC$, let bisectors of $\angle A$, $\angle B$, $\angle C$ intersect circumcircle of $\triangle ABC$ at A_1, B_1 and C_1, respectively. Prove that

$$AA_1 + BB_1 + CC_1 > AB + BC + CA.$$

(Australian Competition in 1982)

Chapter 3 Inequality for the inscribed quadrilateral

Inscribed quadrilateral has not only rich relationships in geometric equalities, but also possesses interesting extremal properties. Since the sides of the inscribed quadrilateral can be represented by the trigonometric functions of the corresponding central angle, this makes it possible to use the trigonometric method to do with the geometric inequalities of the inscribed quadrilateral. The following is such an example.

Example 1. Let $ABCD$ be an inscribed quadrilateral. Prove that

$$|AB - CD| + |AD - BC| \geqslant 2|AC - BD|. \tag{1}$$

(Problem of The 28th Mathematical Olympiad of America)

Proof. Let O be the circumcenter of the inscribed quadrilateral $ABCD$ with radius 1, (see Figure 3.1) $\angle AOB = 2\alpha$, $\angle BOC = 2\beta$, $\angle COD = 2\gamma$, $\angle DOA = 2\delta$, then

$$\alpha + \beta + \gamma + \delta = \pi.$$

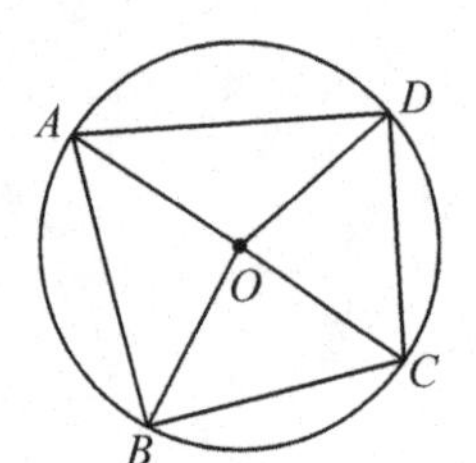

Figure 3.1

Without loss of generality assume that $\alpha \geqslant \gamma$, $\beta \geqslant \delta$, it follows that

$$\begin{aligned} |AB - CD| &= 2|\sin\alpha - \sin\gamma| \\ &= 4\left|\sin\frac{\alpha-\gamma}{2}\cos\frac{\alpha+\gamma}{2}\right| \\ &= 4\left|\sin\frac{\alpha-\gamma}{2}\sin\frac{\beta+\delta}{2}\right|. \end{aligned}$$

Similarly, we have

$$|AD - BC| = 4\left|\sin\frac{\beta-\delta}{2}\sin\frac{\alpha+\gamma}{2}\right|,$$

$$|AC - BD| = 4\left|\sin\frac{\beta-\delta}{2}\sin\frac{\alpha-\gamma}{2}\right|.$$

Therefore

$$\begin{aligned}|AB - CD| - |AC - BD| &= 4\left|\sin\frac{\alpha-\gamma}{2}\right|\left(\left|\sin\frac{\beta+\delta}{2}\right| - \left|\sin\frac{\beta-\delta}{2}\right|\right)\\ &= 4\left|\sin\frac{\alpha-\gamma}{2}\right|\left(\sin\frac{\beta+\delta}{2} - \sin\frac{\beta-\delta}{2}\right)\\ &= 4\left|\sin\frac{\alpha-\gamma}{2}\right|\cdot\left(2\cos\frac{\beta}{2}\cdot\sin\frac{\delta}{2}\right)\\ &\geqslant 0.\end{aligned}$$

Hence

$$|AB - CD| \geqslant |AC - BD|,$$
$$|AD - BC| \geqslant |AC - BD|.$$

Summing up above two inequalities, we obtain inequality (1). □

There is a special inscribed quadrilateral, called double scribed quadrilateral, which means it has both the circumscribed and inscribed circles.

The following example is an inequality of a double scribed quadrilateral. The inequality was found by Mr. Chen Jixian. The following proofs (1) and (2) we introduce were provided by Long Yun (former student of Yali High School of Changsha, China, who was elected to the National Math Winter Campus of China in 1999) and Zhu Qingsan (student), respectively.

Example 2. Let $ABCD$ be a convex double scribed quadrilateral. Denote the radius and area of the circumcirle by R and S, respectively. Let a, b, c, d be the side lengths of the quadrilateral $ABCD$. Prove that

$$abc + abd + acd + bcd \leqslant 2\sqrt{S}(S + 2R^2). \quad (2)$$

Proof 1. We denote the centers of the circumcircle and the inscribed circle of quadrilateral $ABCD$ by O and I, respectively (see Figure 3.2). The tangent points of inscribed circle with sides $AB = a$, $BC = b$, $CD = c$, $DA = d$ are K, L, M, N, respectively. Let $\angle AIN = \angle 1$, $\angle BIK = \angle 2$, $\angle CIL = \angle 3$, $\angle DIM = \angle 4$, and denote $AK = AN = a'$, $BL = BK = b'$, $CL = CM = c'$, $DM = DN = d'$.

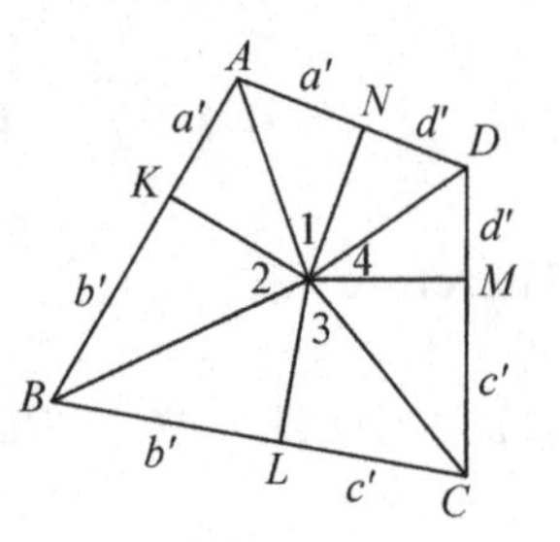

Figure 3.2

Since $ABCD$ has an inscribed circle, we have

$$a + c = b + d.$$

Denote the left side of (1) by H, and without loss of generality, we may assume that the radius of the inscribed circle is 1, it follows that

$$H = (a + c)bd + (b + d)ac = \frac{1}{2}(a + b + c + d)(ac + bd). \quad \text{(a)}$$

Applying

$$a = a' + b',\ b = b' + c',\ c = c' + d',\ d = d' + a',$$

to the right side of (a), yields

$$H = (a' + b' + c' + d')[(a' + b')(c' + d') + (b' + c')(d' + a')]. \quad \text{(b)}$$

Since $\angle A + \angle C = 180°$, we obtain

$$\angle 1 + \angle 3 = 90°.$$

From this, we have $\triangle AIN \backsim \triangle ICL$, hence

$$a'c' = AN \cdot CL = NI \cdot IL = 1. \quad \text{(c)}$$

Similarly

$$b'd' = 1. \quad \text{(d)}$$

Notice that

$$S = \frac{a+b+c+d}{2} \cdot r = a' + b' + c' + d', \tag{e}$$

from equalities (c), (d), (e), we obtain

$$H = S[4 + (a' + c')(b' + d')]. \tag{f}$$

On the other hand, from sine law and $\angle B + 2\angle 2 = 180°$, we have

$$\begin{aligned} R &= \frac{AC}{2\sin\angle B} \\ &= \frac{AC}{2\sin 2\angle 2} \\ &= \frac{AC}{4}\left(\tan\angle 2 + \frac{1}{\tan\angle 2}\right) \\ &= \frac{1}{4}AC(\tan\angle 2 + \tan\angle 4) \\ &= \frac{1}{4}AC \cdot (b' + d'). \end{aligned}$$

Similarly

$$R = \frac{1}{4}BD \cdot (a' + c'),$$

therefore

$$R^2 = \frac{1}{16}AC \cdot BD(a' + c')(b' + d'),$$

but

$$S = \frac{1}{2}AC \cdot BD \cdot \sin\alpha \leqslant \frac{1}{2}AC \cdot BD,$$

where α is the angle of diagonal AC and BD. Hence

$$R^2 \geqslant \frac{1}{8}S(a' + c')(b' + d').$$

From above we have

$$2\sqrt{S}(S + 2R^2) \geqslant 2\sqrt{S}\left(S + \frac{S}{4} + (a' + c')(b' + d')\right)$$

$$= \frac{S^{\frac{3}{2}}}{2}[4 + (a' + c')(b' + d')]. \tag{g}$$

By (f), (g), in order to prove (1), it suffices to prove $\frac{1}{2}S^{\frac{1}{2}} \geqslant 1$, which is equivalent to

$$\sqrt{a' + b' + c' + d'} \geqslant 2. \tag{h}$$

Since $a'c' = 1$, $b'd' = 1$, we obtain

$$a' + c' + b' + d' \geqslant 2\sqrt{a'c'} + 2\sqrt{b'd'} = 4,$$

so we have proved (h). □

The above natural and fluent proof that uses fine triangulation methods won high praise by lots of math olympic masters.

Proof 2. First we prove a lemma.

Lemma 1. In $\triangle ABC$, if $\angle A \geqslant 90^\circ$, then $(b + c)/a \leqslant \sqrt{2}$.

Proof.

$$\frac{b + c}{a} = \frac{\sin B + \sin C}{\sin A} = 2\frac{\sin\frac{B + C}{2}\cos\frac{B - C}{2}}{\sin A}$$

$$\leqslant \frac{2\cos\frac{A}{2}}{2\sin\frac{A}{2}\cos\frac{A}{2}} = \frac{1}{\sin\frac{A}{2}} \leqslant \sqrt{2}.$$ □

Let us prove the original inequality.

Proof. Assuming that the four sides of quadrilateral $ABCD$ be AB, BC, CD, DA (see Figure 3.3), where the four sides lengths are a, b, c, d, respectively. And let the inscribed radius of quadrilateral $ABCD$ be 1. Since

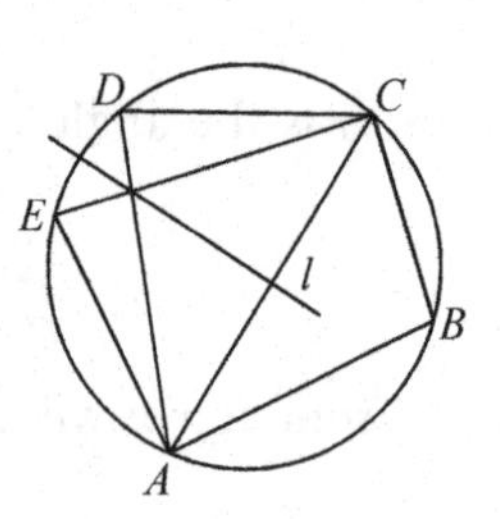

Figure 3.3

$$a + c = b + d = \frac{1}{2}(a + b + c + d) \cdot 1 = S,$$

we get

$$\begin{aligned} H &= abc + abd + acd + bcd \\ &= ac(b + d) + bd(a + c) = (ac + bd)S. \end{aligned} \tag{a}$$

Let the bisector of AC be l, D and E be symmetric to the line l, therefore

$$\triangle ACD \cong \triangle CAE,$$

thus $AE = c$, $CE = d$, and $\angle E = \angle D = \pi - \angle B$, it follows that A, E, C, B lie on a circle, so

$$S = \frac{1}{2}(ac + bd)\sin\alpha, \tag{b}$$

where $\alpha = \angle EAB$.

From (a), (b), we know that the inequality is equivalent to

$$\frac{2S}{2\sin\alpha} \cdot S \leqslant 2\sqrt{S}(S + 2R^2). \tag{c}$$

Notice that $R = \dfrac{BE}{2\sin\alpha}$, so (c) is further equivalent to

$$S^{\frac{3}{2}} \leqslant S\sin\alpha + \frac{BE^2}{2\sin\alpha}, \tag{d}$$

according to the mean value inequality, we get

$$S\sin\alpha + \frac{BE^2}{2\sin\alpha} \geqslant 2\sqrt{\frac{S \cdot BE^2}{2}}.$$

To prove (d), it suffices to prove

$$\sqrt{2}BE \geqslant S. \tag{e}$$

In fact, since $\angle EAB + \angle ECB = 180°$, assuming that $\angle EAB \geqslant 90°$. Applying Lemma 1 to $\triangle ABE$, we get $\dfrac{a + c}{BE} \leqslant \sqrt{2}$, and it follows that

$$\sqrt{BE} \geqslant a + c = S,$$

so we have proved (e). □

The method above used trigonometric function and geometry, and constructed a new inscribed quadrilateral, so transformed the problem.

The inscribed quadrilateral has a famous extremal property: Of all quadrilaterals with given sides, the inscribed quadrilateral has the maximum area.

Theorem 1. Let the four sides of the convex inscribed quadrilateral be a, b, c, d, respectively, and s be half of the perimeter, then the area F of the quadrilateral is

$$F = \sqrt{(s-a)(s-b)(s-c)(s-d)}. \tag{3}$$

If $d = 0$, this is the Heron's area formula for triangular. The proof below quotes from the book-Modern Geometry by Roger A. Johnson. (Translated by Shan Zun, Shanghai Education Publishing House, 1999.)

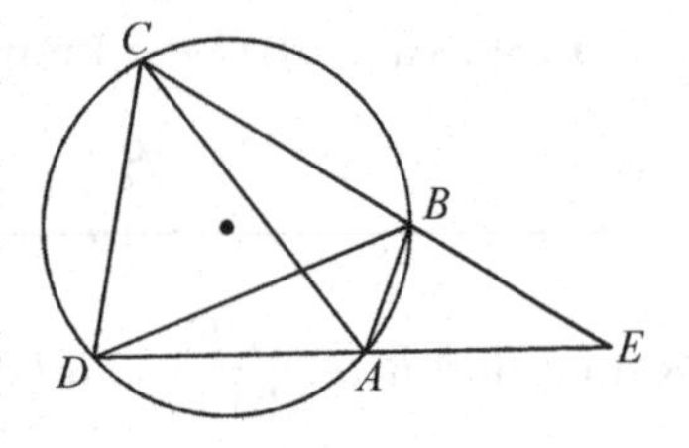

Figure 3.4

Proof. Let the quadrilateral be $ABCD$, and $AB = a$, $BC = B$, $CD = c$, $DA = d$, see Figure 3.4.

If the quadrilateral is a rectangle, the proof is obvious. Otherwise, we assume AD and BC are extended to intersect at point E outside the circle, and $CE = x$, $DE = y$, according to the area formula of the triangle, we obtain the area of $\triangle CDE$

$$S_{\triangle CDE} = \frac{1}{4}\sqrt{(x+y+c)(x+y-c)(x-y+c)(-x+y+c)}. \tag{a}$$

Note that $\triangle ABE \backsim \triangle CDE$, so we have

$$\frac{S_{\triangle ABE}}{S_{\triangle CDE}} = \frac{a^2}{c^2},$$

therefore

$$\frac{F}{S_{\triangle CDE}} = \frac{c^2 - a^2}{c^2}, \tag{b}$$

and from the proportion relations

$$\frac{x}{y} = \frac{y - d}{a},$$

$$\frac{y}{c} = \frac{x - b}{a},$$

we obtain

$$x + y + c = \frac{c}{c - a}(-a + b + c + d),$$

similarly, we can get expressions of $x + y - c$ etc..

Substitute them to (a) and simplifying, we have

$$S_{\triangle CDE} = \frac{c^2}{c^2 - a^2}\sqrt{(s - a)(s - b)(s - c)(s - d)}, \tag{c}$$

substitute (c) into (b), we obtain the inequality (3). □

Generalization. Let the side lengths of a convex quadrilateral be a, b, c and d, respectively, and the sum of opposite angles be $2u$. Prove that the area F can be given by

$$F^2 = (s - a)(s - b)(s - c)(s - d) - abcd\cos^2 u.$$

The proof of it is too tedious and insipid to be given here. But we can see from it that if the four sides of a quadrilateral are given, the inscribed quadrilateral has the maximum area.

The example below uses the extremal property of the inscribed quadrilateral.

Example 3 (Popa's inequality). If the convex quadrilateral with the area F and four sides satisfying $a \leqslant b \leqslant c \leqslant d$.

Prove that

$$F \leqslant \frac{3\sqrt{3}}{4}c^2. \tag{4}$$

Proof. By the extremal property of the inscribed quadrilateral, we only need to prove (4) for inscribed quadrilateral.

$$F^2 = (s-a)(s-b)(s-c)(s-d),$$

where $s = \frac{1}{2}(a+b+c+d)$, $s-d = (a+b+c)-s$. By the arithmetic-geometric inequality, we have

$$\begin{aligned} F^2 &= 3^3\left(\frac{1}{3}s - \frac{1}{3}a\right)\left(\frac{1}{3}s - \frac{1}{3}b\right)\left(\frac{1}{3}s - \frac{1}{3}c\right)(a+b+c-s) \\ &\leqslant 3^3\left[\frac{\left(\frac{1}{3}s - \frac{1}{3}a\right) + \left(\frac{1}{3}s - \frac{1}{3}b\right) + \left(\frac{1}{3}s - \frac{1}{3}c\right) + (a+b+c-s)}{4}\right]^4 \\ &= 3^3\left(\frac{a+b+c}{3\cdot 2}\right)^4 \leqslant 3^3\left(\frac{c}{2}\right)^4, \end{aligned}$$

and the last step uses $a \leqslant b \leqslant c$.

Thus

$$F \leqslant \frac{3\sqrt{3}}{4}c^2.$$

□

The following is another typical problem.

Example 4 (Gaolin's inequality). Let the convex quadrilaterals $ABCD$ and $A'B'C'D'$ have side lengths a, b, c, d and a', b', c', d', and with the area F, F', respectively.

Denote

$$K = 4(ad+bc)(a'd'+b'c') - (a^2-b^2-c^2+d^2)(a'^2-b'^2-c'^2+d'^2).$$

Prove that

$$K \geqslant 16FF'. \tag{5}$$

Proof. By the extremal property of the inscribed quadrilateral, it suffices to consider the inscribed quadrilaterals, see Figure 3.5. Since $\angle B + \angle D = 180°$, we have

$$2F = (ad + bc)\sin B, \qquad \text{(a)}$$

similarly

$$2F' = (a'd' + b'c')\sin B'. \qquad \text{(b)}$$

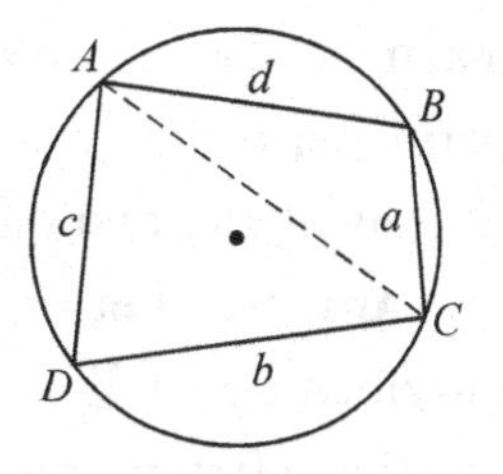

Figure 3.5

On the other hand, by the cosine law, we obtain

$$AC^2 = b^2 + c^2 + 2bc\cos B = a^2 + d^2 - 2ad\cos B,$$

therefore

$$a^2 - b^2 - c^2 + d^2 = 2(ad + bc)\cos B, \qquad \text{(c)}$$

likewise

$$a'^2 - b'^2 - c'^2 + d'^2 = 2(a'd' + b'c')\cos B', \qquad \text{(d)}$$

from relations (a) and (d), we have

$$K - 16FF' = 4(ad + bc)(a'd' + b'c')(1 - \cos(B - B')) \geqslant 0,$$

as desired. □

Remark. By the above proof we can write the inequality even more general

$$0 \leqslant K - 16FF' = 8(ad + bc)(a'd' + b'c'),$$

and the left side of the above inequality is Gaolin's inequality.

Gaolin's inequality can be regarded as the generalization of Neuberg-Pedoe's inequality for quadrilateral.

At the end of this section, we study a much harder extremal problem of the inscribed quadrilateral. The solution is provided by Xiang Zhen (former student of the First High School of Changsha City, China, who won a gold medal at the 44th IMO).

Example 5. For given radius R and area ($S \leqslant 2R^2$) of the circumcircle for a double scribed quadrilateral $ABCD$. Evaluate the maximal value of plm, where p is the semi-perimeter of the quadrilateral, l and m are the lengths of the two diagonals.

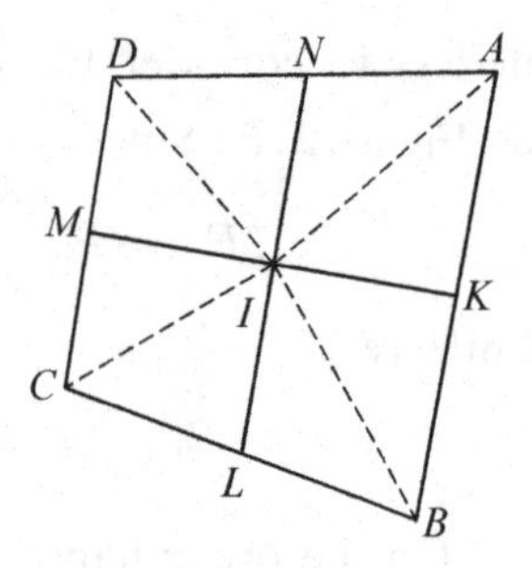

Figure 3.6

Answer. Let r be the radius of the inscribed circle, $\alpha = \angle AIK$, $\beta = \angle BIK$, I be the circumcenter of the quadrilateral $ABCD$, and K be the tangent point of $\odot I$ and line AB, see Figure 3.6.

Since the semi-perimeter of the quadrilateral

$$p = r(\tan\alpha + \cot\alpha + \tan\beta + \cot\beta) = r\left(\frac{2}{\sin 2\alpha} + \frac{2}{\sin\beta}\right),$$

it follows that

$$S = rp = r^2\left(\frac{2}{\sin 2\alpha} + \frac{2}{\sin 2\beta}\right). \qquad \text{(a)}$$

In triangle ABD, we have

$$AB = r(\tan\alpha + \tan\beta),\ AD = r(\tan\alpha + \cot\beta),\ \angle DAB = \pi - 2\alpha.$$

By the cosine law yields

$$\begin{aligned} BD^2 &= r^2[(\tan\alpha + \tan\beta)^2 + (\tan\alpha + \cot\beta)^2 \\ &\quad + 2\cos 2\alpha(\tan\alpha + \tan\beta)(\tan\alpha + \cot\beta)] \\ &= r^2\left(\tan\alpha \cdot \frac{2}{\sin 2\beta} \cdot 4\cos^2\alpha + \frac{4}{\sin^2 2\beta}\right). \end{aligned}$$

Therefore

$$R^2 = \frac{BD^2}{4\sin^2 2\alpha} = r^2\left(\frac{1}{\sin 2\alpha \sin 2\beta} + \frac{1}{\sin^2 2\alpha \sin^2 2\beta}\right). \qquad \text{(b)}$$

Denote $a = \sin 2\alpha$, $b = \sin 2\beta$, we get $a, b \in (0, 1]$, so (a), (b), can be written as

$$S = 2r^2\,\frac{a+b}{ab}, \qquad \text{(c)}$$

$$R^2 = r^2\,\frac{1+ab}{a^2b^2}, \qquad \text{(d)}$$

divided (d) by (c) yields

$$\frac{ab(a+b)}{1+ab}=\frac{S}{2R^2}. \tag{e}$$

(e) is the constraint condition of a, b. By this condition, we deduce the maximal value. Since

$$p=r\cdot\left(\frac{2}{a}+\frac{2}{b}\right),$$
$$lm=4R^2ab$$

we have

$$(plm)^2=64R^4r^2(a+b)^2=16R^2S^2(1+ab),$$

therefore

$$plm=4RS\sqrt{1+ab}. \tag{f}$$

From (e), we obtain

$$\frac{S}{2R^2}=\frac{ab(a+b)}{1+ab}\geqslant\frac{ab\cdot 2\sqrt{ab}}{1+ab}. \tag{g}$$

Denote $\sqrt{ab}=x$, with $x\in(0, 1]$, (g) gives

$$4R^2\cdot x^3-S\cdot x^2-S\leqslant 0. \tag{h}$$

Define function $f(x)=4R^2\cdot x^3-S\cdot x^2-S$, notice that

$$f(0)=-S\leqslant 0,\ f(1)=4R^2-2S\geqslant 0$$

and

$$\begin{cases} f'(x)\geqslant 0,\ x\geqslant\dfrac{S}{6R^2}, \\ f'(x)<0,\ 0<x<\dfrac{S}{6R^2}. \end{cases}$$

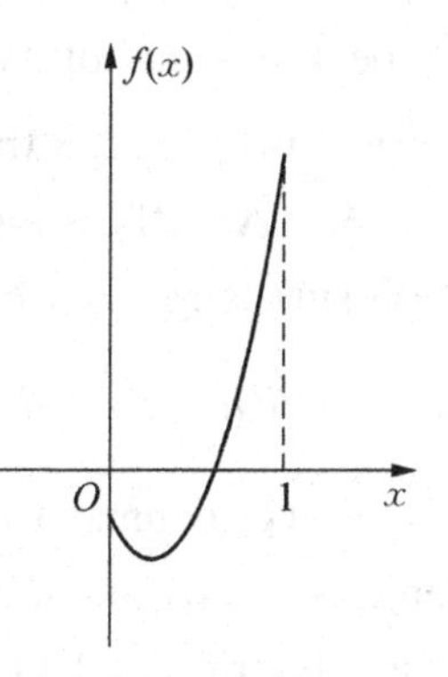

Figure 3.7

Thus $f(x)$ decreasing first and then increasing. Therefore $f(x)$ has a unique root in $(0, 1)$, see Figure 3.7.

By (h) we see that $f(x)\leqslant 0$, therefore,

$$\sqrt{ab} \leqslant t,$$

then $ab \leqslant t^2$. Applying it to (f), we obtain

$$plm = 4RS\sqrt{1+ab} \leqslant 4RS\sqrt{1+t^2}.$$

If $a = b = t$, the equality holds, thus the maximal of plm is $4RS\sqrt{1+t^2}$, and t is the root of $4R^2x^3 - Sx^2 - S = 0$ in interval $(0, 1]$.

Exercises 3

1. Let $ABCD$ be an inscribed convex quadrilateral with interior angles and exterior angles no less than 60°. Prove that

$$\frac{1}{3}\,|AB^3 - AD^3| \leqslant |BC^3 - CD^3| \leqslant 3\,|AB^3 - AD^3|,$$

and point out the condition such that the equality holds.

2. Let $ABCD$ be a convex circumscribed quadrilateral $ABCD$ with area S and the circumcenter is inside the quadrilateral. The intersection point of two diagonals is denoted by E, and let M, N, P, Q be the projection of E on four sides, respectively. Prove that the area of $MNPQ$ is no more than $\frac{S}{2}$.

3. Let a, b, c and d and a', b', c', d' be the side lengths, S and S' be the areas of two convex quadrilaterals $ABCD$ and $A'B'C'D'$, respectively. Prove that $aa' + bb' + cc' + dd' \geqslant 4\sqrt{SS'}$.

4. (An Zhenping) Let $ABCD$ be an circumscribed quadrilateral with side lengths a, b, c, d. Prove that

$$a^2b(a-b) + b^2c(b-c) + c^2d(c-d) + d^2a(d-a) \geqslant 0.$$

5. (Groenman) Let $ABCD$ be an inscribed quadrilateral with side lengths a, b, c, d. And ρ_a is the radius of the circle outside the quadrilateral, and tangent to the edges AB, CB, and extended line

DA. The ρ_b, ρ_c, ρ_d are defined similarly. Prove that

$$\frac{1}{\rho_a}+\frac{1}{\rho_b}+\frac{1}{\rho_c}+\frac{1}{\rho_d}\geqslant\frac{8}{\sqrt[4]{abcd}}$$

and the equality holds if and only if the $ABCD$ is a square.

Chapter 4 The area inequality for special polygons

The area inequalities and extreme value problem for polygons have attracted much attention.

Some area inequalities for special polygons such as parallelogram and triangle often appeared at middle school math competitions. In this section, we introduce some interesting results, and try our best to treat the area problems more generally.

First, we look into the relationship of area between the parallelogram and the inscribed triangle in it. A well-known conclusion of it is: any area of the inscribed triangle does not exceed the half of the parallelogram area.

The proof of this conclusion is quite simple, see Figure 4. 1, just make line passing point Q, the apex of the triangle PQR, paralleling to AB, and consider the relations between the areas of the small parallelogram and the triangle it contains.

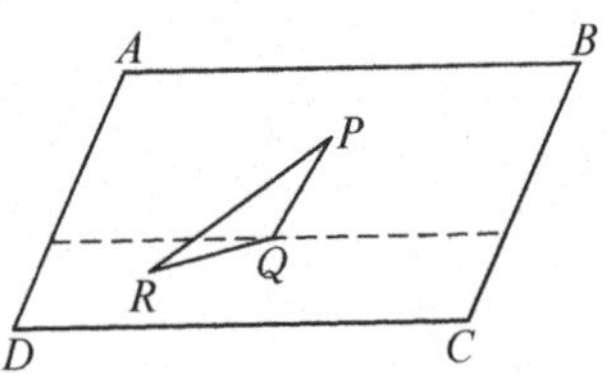

Figure 4. 1

Now consider the opposite problem: what is the relationship of the areas between the parallelogram and the triangle within? The answer is a useful theorem as follows.

Theorem 1. The area of a parallelogram in any triangle is no more than half area of the triangle.

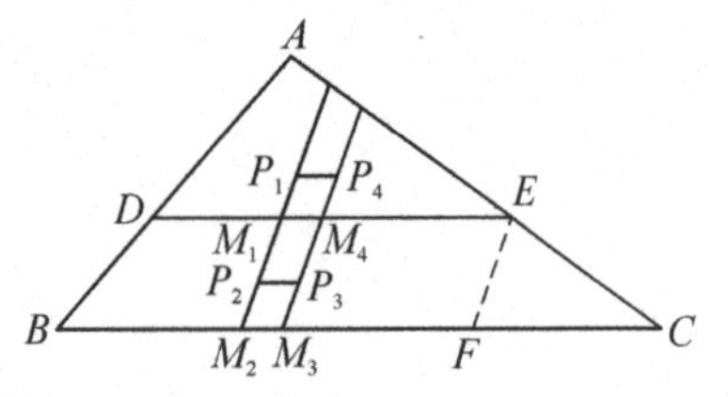

Figure 4. 2

Proof. Considering the parallelogram

$P_1P_2P_3P_4$ contained in a triangle ABC, see Figure 4.2. Let M_2, M_3 be the intersection points of lines P_1P_2, P_3P_4 with BC, respectively. Assume that $M_2M_1 = P_2P_1$, $M_3M_4 = P_3P_4$, and $M_1 \in P_1P_2$, $M_4 \in P_3P_4$, so $M_1M_2M_3M_4$ is a parallelogram, and

$$S(M_1M_2M_3M_4) = S(P_1P_2P_3P_4).$$

Let the intersection point of the line M_1M_4 with AB and AC be D and E, respectively. Let EF be parallel to AB, so that $BDEF$ is a parallelogram, hence

$$S(BDEF) \geqslant S(M_1M_2M_3M_4) = S(P_1P_2P_3P_4).$$

If we want to prove

$$S(P_1P_2P_3P_4) \leqslant \frac{1}{2}S_{\triangle ABC},$$

we must prove

$$S(BDEF) \leqslant \frac{1}{2}S_{\triangle ABC}. \tag{a}$$

Now we proceed to prove (a).

Denote $\lambda = \frac{AD}{AB}$, see Figure 4.3. Since

$$\triangle ADE \backsim \triangle ABC,$$

we obtain

$$S_{\triangle ADE} = \lambda^2 S_{\triangle ABC}.$$

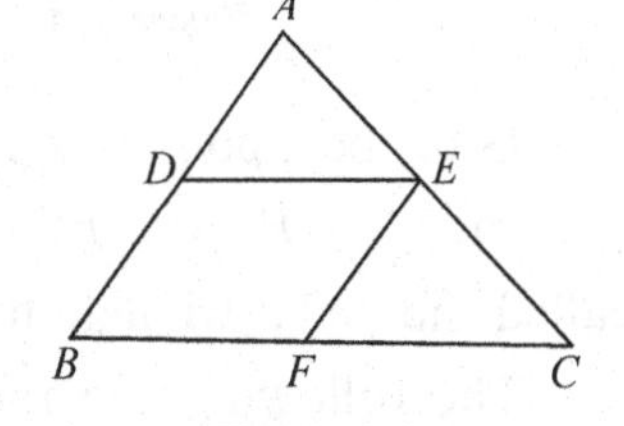

Figure 4.3

Similarly,

$$S_{\triangle EFC} = (1-\lambda)^2 S_{\triangle ABC}.$$

Therefore,

$$S_{\triangle ADE} + S_{\triangle EFC} = [\lambda^2 + (1-\lambda)^2]S_{\triangle ABC} \geqslant \frac{1}{2}S_{\triangle ABC}.$$

Thus

$$S(BDEF) = S_{\triangle ABC} - (S_{\triangle ADE} + S_{\triangle EFC}) \leqslant \frac{1}{2} S_{\triangle ABC},$$

as desired, and equality holds if and only if D, E, F are the midpoints. □

The method above is a typical method of transformation, that is to say, the parallelogram $P_1P_2P_3P_4$ is transformed into parallelogram $M_1M_2M_3M_4$ of which a pare sides are parallel to the side BC, and then transformed into a special parallelogram $BDEF$ whose two pare sides are parallel to the sides of triangle respectively, thus the problem has been greatly simplified.

Let P be a point in $\triangle ABC$, see Figure 4.4, and D, E, F be the intersection points of lines AP, BP, CP with the three sides, respectively. The $\triangle DEF$ is called Ceva's triangle to P.

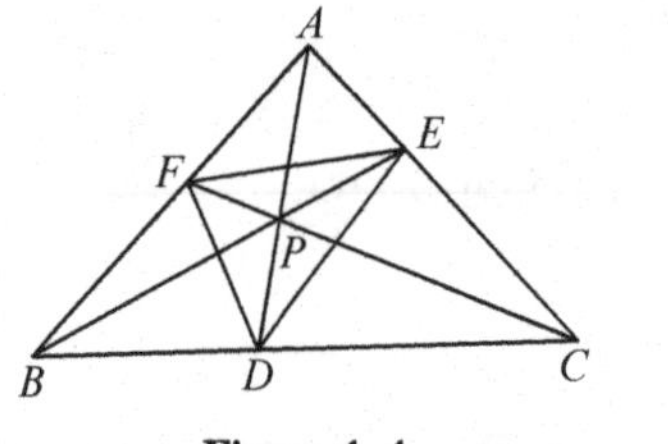

Figure 4.4

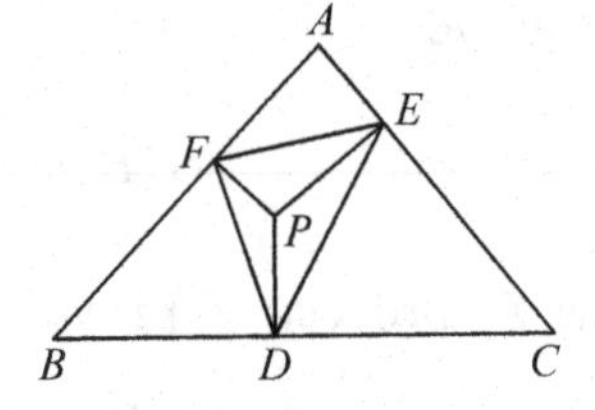

Figure 4.5

Let P be a point in $\triangle ABC$, see Figure 4.5, and D, E and F be the projection of P onto BC, CA and AB, respectively. The $\triangle DEF$ is called the pedal triangle to P.

The following are famous theorems on Ceva's triangle and the pedal triangle.

Proposition 7. Let P be the point in $\triangle ABC$, then the area of Ceva's triangle $\triangle DEF$ to P is not more than $\frac{1}{4} S_{\triangle ABC}$.

Proposition 8. Let P be the point in $\triangle ABC$, then the area of triangle of pedal triangle $\triangle DEF$ to P is not more than $\frac{1}{4} S_{\triangle ABC}$.

Mr. Yang Lin noticed the relationship between Theorem 1 and Proposition 7, and found that the Proposition 7 is the corollary of Theorem 1 by the expansion of Ceva's $\triangle ABC$ as follows.

Example 1. Let P be the point in $\triangle ABC$, and $\triangle DEF$ be Ceva's triangle to P. Show that in $\triangle ABC$ there is a parallelogram with two sides of $\triangle DEF$ as its adjacent sides.

Proof. Let G be the orthocenter of $\triangle ABC$, N, M be the midpoints of the sides AC and AB, respectively, see Figure 4.6. Without loss of generality, we assume that P is at the side or interior of $ANGM$, so E, F lie on segment AN, AM or on endpoints of them, and

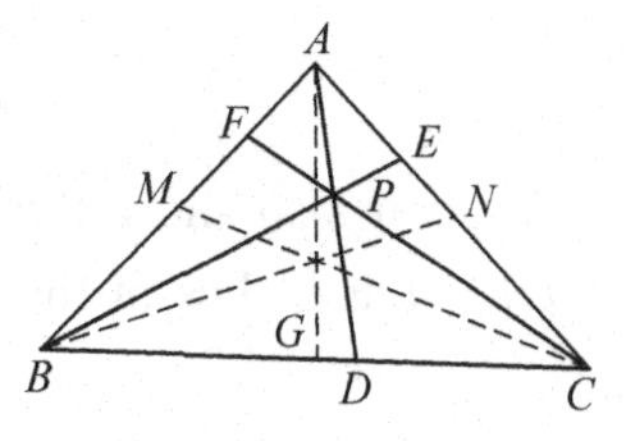

Figure 4.6

$$\frac{AF}{FB} \leqslant 1, \frac{AE}{EC} \leqslant 1,$$

and without loss of generality, we assume that

$$\frac{AF}{FB} \leqslant \frac{AE}{EC}.$$

By Ceva's theorem, we obtain

$$\frac{AF}{FB} \cdot \frac{BD}{DC} \cdot \frac{CE}{EA} = 1,$$

therefore

$$\frac{BD}{DC} = \frac{AE}{CE} \cdot \frac{FB}{AF} \geqslant 1.$$

We can construct the parallelogram $FEDE'$ with adjacent sides EF and ED, see Figure 4.7. So we need only to show that E' lies in the interior or on the side of $\triangle ABC$.

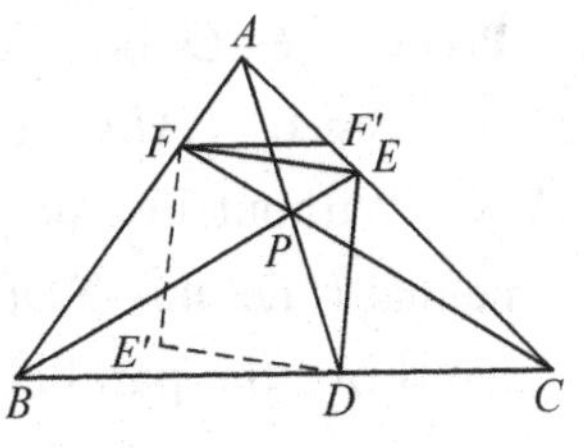

Figure 4.7

Draw a line FF' parallel to BC, so F' lies on AC. And since

$$\frac{AF}{FB} \leqslant \frac{AE}{EC},$$

it follows that F' lies on segment AE or endpoints. Because

$$\angle E'DF = \angle EDF \leqslant \angle F'FD = \angle FDB,$$

we see that DE' lies in the interior of $\angle FDB$.

Likewise

$$\frac{CE}{EA} \geqslant 1 \geqslant \frac{CD}{DB},$$

and we can also show that FE' lies in the interior or on the side of $\angle BDF$, thus E' lies in the interior of $\triangle FDB$ as desired. □

Theorem 1 and Proposition 7 are linked by Example 1, that is to say

$$\text{Theorem 1} \Rightarrow \text{Proposition 7}.$$

A natural question is, does the pedal triangle of the inner point P have a similar extension property as Ceva's triangle?

It is easy to see that the pedal triangle about the inner point in obtuse triangle does not have the extension property generally, but the answer is positive to the acute triangle.

Example 2. Let P be the interior point of acute triangle $\triangle ABC$, $\triangle DEF$ be the pedal triangle about P. Show that in $\triangle ABC$ there is a parallelogram with two sides of $\triangle DEF$ as its adjacent sides.

Proof. Let O be the circumcenter of $\triangle ABC$. Since $\triangle ABC$ is acute, O lies in $\triangle ABC$. Without loss of generality we may assume that P lies in $\triangle AOB$, see Figure 4.8.

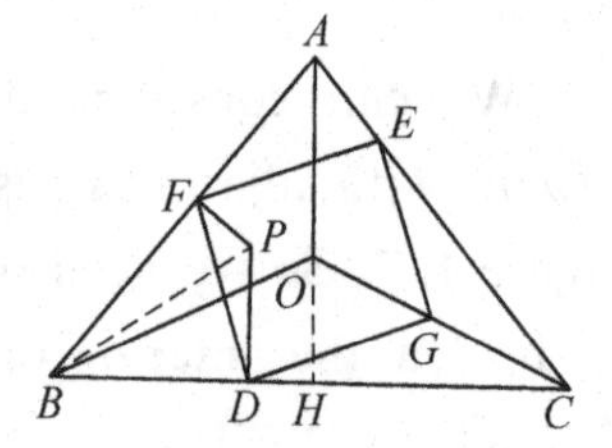

Figure 4.8

To prove the parallelogram $DFEG$ with

adjacent sides FE, FD lies in $\triangle ABC$, we need only to prove

$$\angle FEG \leqslant \angle FEG, \tag{a}$$

$$\angle FDG \leqslant \angle FDC. \tag{b}$$

Since

$$\angle FEG = \angle AFE + \angle BFD,$$
$$\angle FEC = \angle AFE + \angle A,$$

so to prove (a), we need only to prove

$$\angle BFD \leqslant \angle A. \tag{c}$$

In fact, since four points B, F, P and D are on a circle, we have

$$\angle BFD = \angle BPD. \tag{d}$$

Draw line $OH \perp BC$ with H the foot of perpendicular, and by

$$\angle PBD \geqslant \angle OBH,$$

we have

$$\angle BPD \leqslant \angle BOH. \tag{e}$$

And since O is the circumcenter of $\triangle ABC$, it follows that

$$\angle BOH = \angle BAC = \angle A. \tag{f}$$

From (d), (e), (f), (c) the proof of (a) is complete. The proof of (b) is similar. □

Remark. By Example 2, we know that for an acute triangle, one can induct the Proposition 8 from Theorem 1. Example 2 was a problem in the second Olympic Test of western China. (In order to reduce the difficulty, the author set the condition that P lies in the $\triangle ABC$.)

In the following the topic is about the five-points inside a triangle. The question is provided by A. Soifer to the Mathematical Olympiad at Colorado. He raised and proved that for arbitrary five points in a triangle with unit area, there is at least one triangle formed by three points of five, its area is not more than $\frac{1}{4}$.

It is not difficult to prove the points cannot be less than five, but focus on the number of triangles, we have been able to improve the conclusions of the problem. The following example is first discovered and proved by Mr. Huang Renshou.

Example 3. Any given five points in unit area triangle, there exists two different groups of three points, and the area of triangle formed by them is not more than $\frac{1}{4}$.

We must use the following lemma.

Lemma 1. Let the convex quadrilateral be in the unit area triangle, and there exist a triangle formed by three vertexes of the convex parallelogram, the area of which is not more than $\frac{1}{4}$.

Proof. Since four vertexes of convex quadrilateral can be both on the sides of one triangle, so Lemma 1 is essentially the well-known the second question in the first winter math camp. Let P_1, P_2, P_3, P_4 be on the sides of the $\triangle ABC$. Prove that there exists a triangle the area of which is less than or equal to $\frac{1}{4}$ among $\triangle P_1P_2P_3$, $\triangle P_1P_3P_4$, $\triangle P_1P_2P_4$, $\triangle P_2P_3P_4$. □

We prove Example 3 as follows.

Proof. If the convex hull of the five points is a segment, the conclusion is confirmed.

If the convex hull of five points is a triangle, we have the five non-intersecting triangles which can be made by these fives points, and the sum of these triangles area is less than or equal to 1, so there exists two triangles, the area of which is not more than $\frac{1}{4}$, see Figure 4.9.

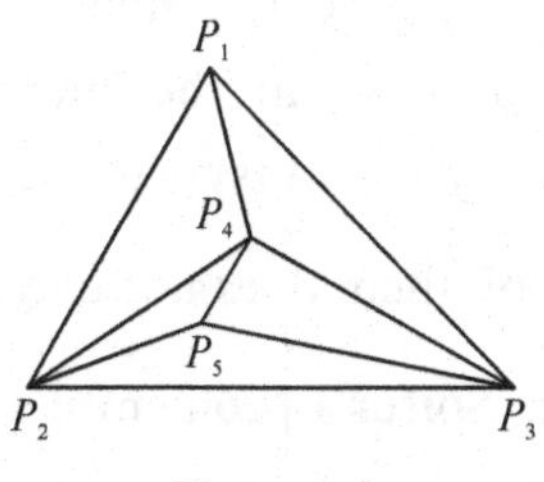

Figure 4.9

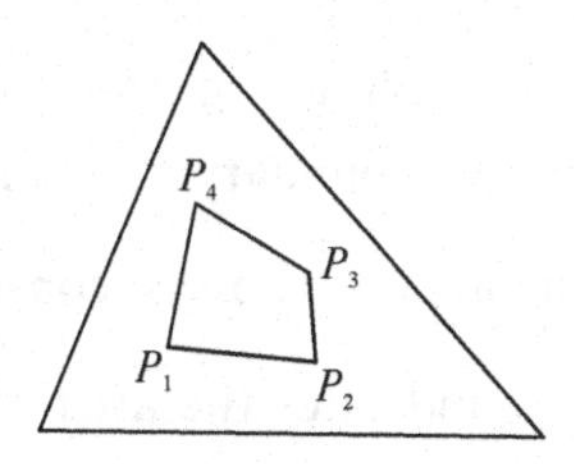

Figure 4.10

If the convex hull of five points is a convex quadrilateral, without loss of generality, let the five points distribute as the Figure 4.10, that is to say P_5 lies in the convex quadrilateral $P_1P_2P_3P_4$. By Lemma 3, it follows that there exists a triangle formed by three points among P_1, P_2, P_3, P_4, the area of which is not more than $\frac{1}{4}$. And since

$$S_{\triangle P_1P_2P_5} + S_{\triangle P_2P_3P_5} + S_{\triangle P_3P_4P_5} + S_{\triangle P_4P_1P_5} \leqslant S(P_1P_2P_3P_4) \leqslant S_{\triangle ABC} = 1,$$

it follows that there exists one triangle among $S_{\triangle P_1P_2P_5}$, $S_{\triangle P_2P_3P_5}$, $S_{\triangle P_3P_4P_5}$, $S_{\triangle P_4P_1P_5}$, the area of which is less than or equal to $\frac{1}{4}$.

If the convex hull of five points is convex pentagon, so the convex quadrilateral which is formed by arbitrarily four points is in $\triangle ABC$ as Figure 4.11, so the number of the convex quadrilateral like this is $C_5^4 = 5$. There are five triangles (including count twice) the area of which is less than $\frac{1}{4}$, and every triangle is counted twice at most, so the number of triangles which area is no more than $\frac{1}{4}$ is greater than 2. □

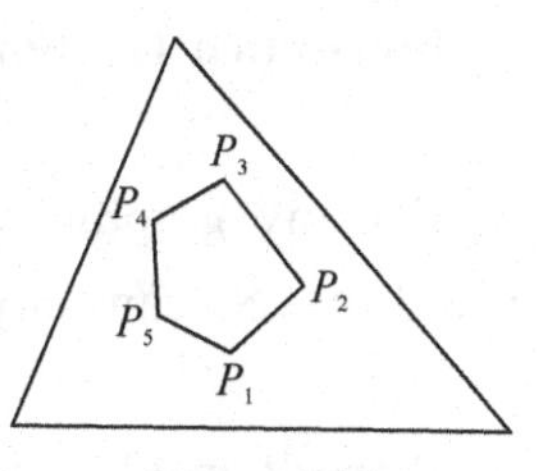

Figure 4.11

Remark. 1. We can prove that the conclusion above can be improved to: There are three (but not four) triangles of which each area is no more than $\frac{1}{4}$. The proof is too long to be given here.

2. Given an arbitrary graph F, let S_F be the smallest positive integer n satisfying the following conditions, in the internal of F (including the boundary), given arbitrary n points, there exist three points, the area of triangle constituted by them does not exceed $\frac{|F|}{4}$, where $|F|$ indicates the area of F. A. Soifer's problem is equivalent to the following proposition.

Proposition 9. For any triangle F, $S_F = 5$.

A. Soifer further proved the following:

Proposition 10. For any parallelogram F, $S_F = 5$.

A natural question is this: whether $S_F = 5$ holds for any of the graphics F or not?

The answer is negative, A. Soifer proved the following:

Proposition 11. For regular pentagon, $S_F = 6$.

For any graphics F, what value can S_F attain? A. Soifer had proved that S_F can only take in a very small range.

Proposition 12. For convex graphics, $4 \leqslant S_F \leqslant 6$.

The further improvements of Proposition 12 are as follows:

Proposition 13. For convex graphics F, $S_F \neq 4$.

Proposition 14. For convex graphics F, $S_F = 5$, or $S_F = 6$.

However, an interesting open question is: What kind of convex graphics F such that $S_F = 5$, and what kind of convex graphics F such that $S_F = 6$?

The following discussion is about what kind of parallelogram or triangle can cover the convex polygon with area 1. We have:

Example 4. (1) The convex polygon with area 1 can be covered by parallelogram with area 2. (2) The convex polygon with area 1 can be covered by triangle with area 2.

Proof. (1) First let a convex polygon M with area 1 be on one side of the support line AB, there exists one point C in M with the greatest distance to AB, C may be a vertex of M or lies in the line parallel to AB. Connect AC, see Figure 4.12, and it divides M into two parts M_1, M_2 (if AC is a side of M, there is no M_1, or M_2). Assume that points D_1 and D_2 are the farthest points to line AC on M, and located at two sides of AC. Draw a straight line parallel to AB through C, and straight lines l_1, l_2 parallel to AC, so straight lines AB, l, l_1 and l_2 constitute a parallelogram P containing M.

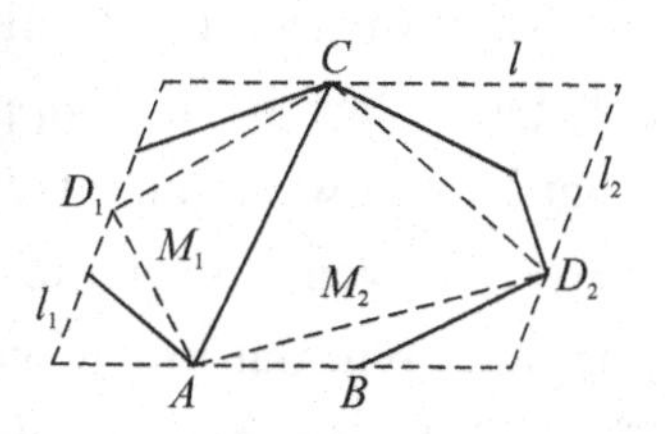

Figure 4.12

Since M_1 and M_2 are convex, they contain $\triangle AD_1C$, $\triangle AD_2C$.

Suppose that P divided by AC into two parallelograms P_1 and P_2, so

$$S_{\triangle AD_1C} = \frac{1}{2}S(P_1),\ S_{\triangle AD_2C} = \frac{1}{2}S(P_2)$$

with $S(X)$ being the area of X, therefore

$$\begin{aligned} S(P) &= S(P_1) + S(P_2) \\ &= 2S_{\triangle AD_1C} + 2S_{\triangle AD_2C} \\ &\leqslant 2S(M_1) + 2S(M_2) = 2S(M) = 2 \end{aligned}$$

as desired.

(2) Let u be a given polygon with area 1, now we consider the internal triangle $\triangle A_1A_2A_3$ with largest area. We discuss it in two cases.

(a) If $\triangle A_1A_2A_3 \leqslant 1/2$. See Figure 4.13, draw three straight lines passing vertexes of $\triangle A_1A_2A_3$ and parallel to the opposite sides, respectively. These three lines constitute a triangle, denote by T, so the area of T is 2.

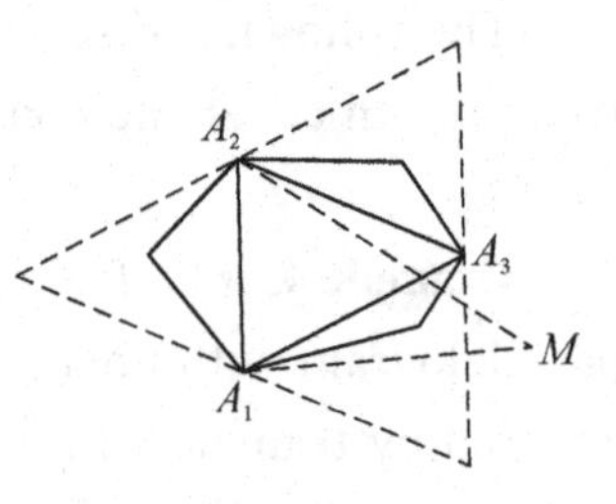

Figure 4.13

Therefore we need only to prove that polygon u lies in T. Suppose that some point M of u were out of T, then the distance of M to some side of $\triangle A_1A_2A_3$ would be greater than that of the vertex A_3 of $\triangle A_1A_2A_3$ to this side. Without loss of generality, assume this side were A_1A_2, see Figure 4.13, in this case the area of $\triangle A_1A_2M$ would be greater than that of $\triangle A_1A_2A_3$. That is contrary to the fact that $\triangle A_1A_2A_3$ has the maximum area in u.

(b) If $\triangle A_1A_2A_3 > 1/2$. There are three parts of u but outside of $\triangle A_1A_2A_3$, see Figure 4.14. In each part, construct triangles with the largest area and one side of $\triangle A_1A_2A_3$ as the base. Denote these triangles by $\triangle B_1A_2A_3$, $\triangle B_2A_1A_3$ and $\triangle B_3A_1A_2$, respectively, then draw lines through B_1, B_2 and B_3 and parallel to A_2A_3, A_1A_3 and A_1A_2, respectively. Thus, we obtain a larger triangle $\triangle C_1C_2C_3$, denoted by C. We can prove that u lies in triangle C as case (a).

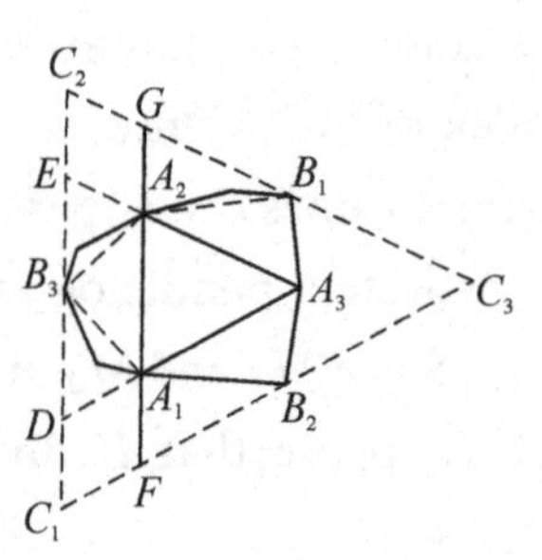

Figure 4.14

Note that u is a convex polygon, then

$$S(A_1B_3A_2B_1A_3B_2) \leqslant S(u) = 1.$$

So we need only to prove that

$$S_{\triangle C_1C_2C_3} \leqslant 2S(A_1B_3A_2B_1A_3B_2). \tag{1}$$

Since $\triangle C_1C_2C_3 \backsim \triangle A_1A_2A_3$, in order to calculate the area of $\triangle C_1C_2C_3$, we denote

$$\frac{S_{\triangle A_1A_2B_3}}{S_{\triangle A_1A_2A_3}} = \lambda_3, \frac{S_{\triangle A_1A_3B_2}}{S_{\triangle A_1A_2A_3}} = \lambda_2, \frac{S_{\triangle A_2A_3B_1}}{S_{\triangle A_1A_2A_3}} = \lambda_1,$$

it follows that

$$\frac{S_{\triangle C_1C_2C_3}}{S_{\triangle A_1A_2A_3}} = (\lambda_1 + \lambda_2 + \lambda_3 + 1)^2. \tag{2}$$

By hypothesis $S_{\triangle A_1A_2B_3} \geqslant 1/2$, we have

$$\begin{aligned} \lambda_1 + \lambda_2 + \lambda_3 &= \frac{S_{\triangle A_1A_2B_3} + S_{\triangle A_1A_3B_2} + S_{\triangle A_2A_3B_1}}{S_{\triangle A_1A_2A_3}} \\ &\leqslant \frac{S(u) - S_{\triangle A_1A_2A_3}}{S_{\triangle A_1A_2A_3}} \\ &= \frac{1 - S_{\triangle A_1A_2A_3}}{S_{\triangle A_1A_2A_3}} \\ &< 1, \end{aligned} \tag{3}$$

and

$$\begin{aligned} \frac{S(A_1B_3A_2B_1A_3B_2)}{S_{\triangle A_1A_2A_3}} &= \frac{S_{\triangle A_1A_2A_3} + S_{\triangle B_1A_2A_3} + S_{\triangle B_2A_1A_3} + S_{\triangle B_3A_1A_2}}{S_{\triangle A_1A_2A_3}} \\ &= \lambda_1 + \lambda_2 + \lambda_3 + 1. \end{aligned} \tag{4}$$

By (2), (3), (4), we obtain

$$\frac{S_{\triangle C_1C_2C_3}}{S(A_1B_3A_2B_1A_3B_2)} = \lambda_1 + \lambda_2 + \lambda_3 + 1 < 2.$$

Thus (1) holds, as desired. □

Now we consider another interesting question: what is the largest area of a triangle inscribed in a convex polygon with area 1? The following examples partly give the answer.

Example 5. (1) Let M be a convex polygon with area 1, and l be an arbitrary given line. Prove that there exists a triangle inscribed in M with one side parallel to l and area greater than or equal to $\frac{3}{8}$.

(2) Let M be a regular hexagon with area 1, and l be an arbitrary given line. Prove that there does not exist inscribed triangle in M with one side parallel to l and area greater than $\frac{3}{8}$.

Proof. (1) As Figure 4.15 shows, draw two supporting lines l_1, l_2 of M parallel to l so that M lies in the zonal region and the vertexes A and B on the parallel lines. Let the width between l_1 and l_2 be d, draw three straight lines l'_1, l_0, l'_2, divide the zonal region into four small strips with the same width $\frac{1}{4}d$. Assume the boundary of M intersect l'_1 at points P and Q, and intersect l'_2 at points R and S. (Since M is convex, its side cannot entirely lie on a straight line.)

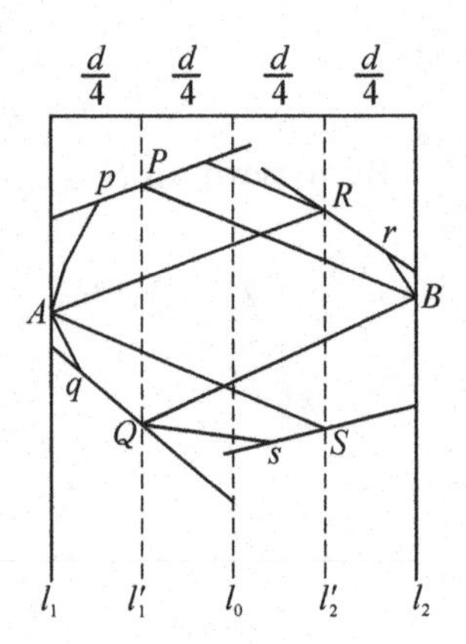

Figure 4.15

Denote p the line on which the side of M passing the point P lies, (if P is the vertex, we can choose any adjacent side). Denote r and s similarly. The trapezoid constructed by lines p, q, l_0 and l_1 has area $\frac{d}{2} \cdot RQ$. Similarly, the trapezoid constructed by lines r, r, l_0 and l_2 has area $\frac{d}{2} \cdot RS$. Since the union set of T_1 and T_2 contains M, so we have

$$S(M) \leqslant S(T_1) + S(T_2) = \frac{d}{2} \cdot PQ + \frac{d}{2} \cdot RS = \frac{d}{2} PQ + RS.$$

Now we consider two triangles $\triangle ARS$ and $\triangle BPQ$, and we find that they are both triangles inscribed in M, and

$$S_{\triangle ARS} = \frac{1}{2} \cdot RS \cdot \frac{3}{4}d, \; S_{\triangle BPQ} = \frac{1}{2} \cdot PQ \cdot \frac{3}{4}d,$$

therefore

$$S_{\triangle ARS} + S_{\triangle BPQ} = (PQ + RS) \cdot \frac{3}{8}d = \frac{3}{4}(PQ + RS) \cdot \frac{1}{2}d$$
$$\geqslant \frac{3}{4}S(M) = \frac{3}{4},$$

so at least one of the following inequalities holds:

$$S_{\triangle ARS} \geqslant \frac{3}{8}, \ S_{\triangle BPQ} \geqslant \frac{3}{8}.$$

(2) Let M be a regular hexagon $ABCDEF$, and $l \parallel AB$, see Figure 4.16. Let $\triangle PQR$ have the largest area inscribed in M, and $PQ \parallel AB$. Without loss of generality, we assume that P and Q be in FA and BC, respectively, it follows that R must be in DE. Let the sides of regular hexagon equal one, and write $AP = BQ = a$, so we obtain

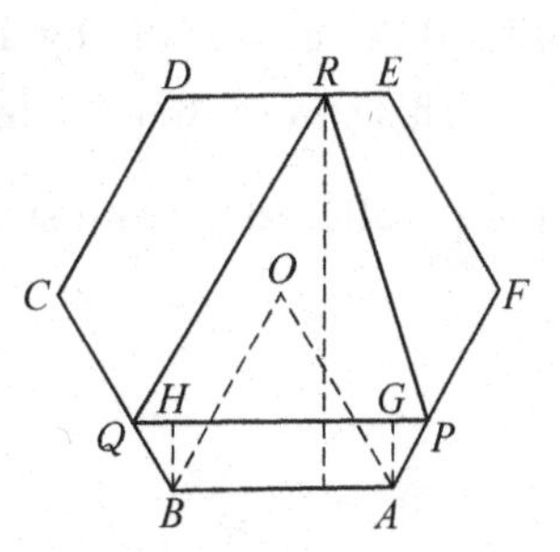

Figure 4.16

$$PQ = AB + PG + QH$$
$$= 1 + \frac{a}{2} + \frac{a}{2} = 1 + a,$$

and

$$h(PQR) = RS - AG = \sqrt{3} - \frac{a\sqrt{3}}{2}$$
$$= (2 - a)\frac{\sqrt{3}}{2},$$

thus

$$S_{\triangle PQR} = \frac{1}{2}(1 + a)(2 - a)\frac{\sqrt{3}}{2}$$
$$= \frac{\sqrt{3}}{4}(2 + a - a^2)$$
$$= \frac{\sqrt{3}}{4}\left(2 + \frac{1}{4} - \left(a - \frac{1}{2}\right)^2\right).$$

From this we see that if $a = 1/2$, the area of $S_{\triangle PQR}$ is the largest, and

$$(S_{\triangle PQR})_{\max} = \frac{9\sqrt{13}}{16},$$

but the area of regular hexagon is

$$6S(OAB) = 6 \cdot \frac{\sqrt{3}}{4} = \frac{3\sqrt{3}}{2},$$

with O the center of regular hexagon.

This shows that the largest area of triangle inscribed in M with one side parallel to a given straight line l is $\frac{3}{8}S(M)$, so the claim holds. □

Exercises 4

1. Let the circumradius of an obtuse $\triangle ABC$ be 1. Prove that $\triangle ABC$ can be covered by an isosceles triangle with hypotenuse length $\sqrt{2}+1$.

2. If a convex polygon M cannot cover any triangle with area 1, prove that M can be covered by a triangle with area 4.

3. (Li Shijie) Let D, E, F be points on sides BC, CA, AB of $\triangle ABC$ respectively, different from vertexes A, B, C. Denote the area of $\triangle ABC$, $\triangle AEF$, $\triangle BDF$, $\triangle CDE$, $\triangle DEF$, by S, S_1, S_2, S_3, S_0, respectively. Prove that

$$S_0 \geqslant 2\sqrt{\frac{S_1 S_2 S_3}{S}},$$

the equality holds if and only if AD, BE, CF intersect at a point in $\triangle ABC$.

4. Show that, it is impossible to put two non-overlapping squares with side length more than $\sqrt{\frac{2}{3}}$ in a square with side length one.

5. Given any n points on plane, and any three of them can be formed a triangle. Let u_n be the ratio of largest area to the smallest area of the triangles, find the minimal value of u_5.

Chapter 5 Linear geometric inequalities

Many linear geometric inequalities give us the impression: simple but unusual, easy to be remembered. The proof of them is either ordinary or difficult. Most linear geometric inequalities in math contests are full of challenge.

Erdös-Mordell's inequality is the most famous one of linear geometric inequalities which we introduce here first.

Example 1 (Erdös-Mordell's inequality). Suppose that point P is in $\triangle ABC$. Let $PD = p$, $PE = q$, and $PF = r$ be distances from P to the sides BC, CA, and AB, respectively. Let $PA = x$, $PB = y$, $PC = z$, then

$$x + y + z \geqslant 2(p + q + r). \tag{1}$$

The equality holds if and only if $\triangle ABC$ is equilateral and P is its center.

The following are five proofs to the inequality. Proof 1 is simple and widely cited, which was given by L. J. Mordell in 1937.

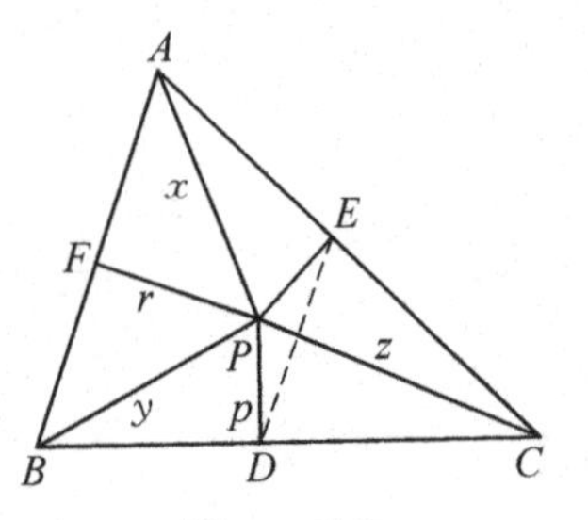

Figure 5.1

Proof. 1. Since $\angle DPE = 180° - \angle C$ (see Figure 5.1), by the cosine law, we get

$$\begin{aligned} DE &= \sqrt{p^2 + q^2 + 2pq\cos C} \\ &= \sqrt{p^2 + q^2 + 2pq\sin A\sin B - 2pq\cos A\cos B} \end{aligned}$$

$$= \sqrt{(p\sin B + q\sin A)^2 + (p\cos B - q\cos A)^2}$$
$$\geqslant \sqrt{(p\sin B + q\sin A)^2}$$
$$= p\sin B + q\sin A.$$

Since P, D, C, E are on a circle, the line segment CP is the diameter of the circle, so

$$z = \frac{DE}{\sin C} \geqslant \left(\frac{\sin B}{\sin C}\right)p + \left(\frac{\sin A}{\sin C}\right)q,$$

similarly,

$$x \geqslant \left(\frac{\sin B}{\sin A}\right)r + \left(\frac{\sin C}{\sin A}\right)q,$$
$$y \geqslant \left(\frac{\sin A}{\sin B}\right)r + \left(\frac{\sin C}{\sin B}\right)p.$$

Adding above three inequalities together, we get

$$x + y + z \geqslant \left(\frac{\sin B}{\sin C} + \frac{\sin C}{\sin B}\right)p + \left(\frac{\sin A}{\sin C} + \frac{\sin C}{\sin A}\right)q + \left(\frac{\sin B}{\sin A} + \frac{\sin A}{\sin B}\right)r$$
$$\geqslant 2(p + q + r).$$

□

The following Proof 2 is given by Mr. Zhang Jingzhong, who applied the method of area subtly and concisely.

Proof. 2. Make MN through P such that $\angle AMN = \angle ACB$, then $\triangle AMN \backsim \triangle ACB$. (See Figure 5.2.)

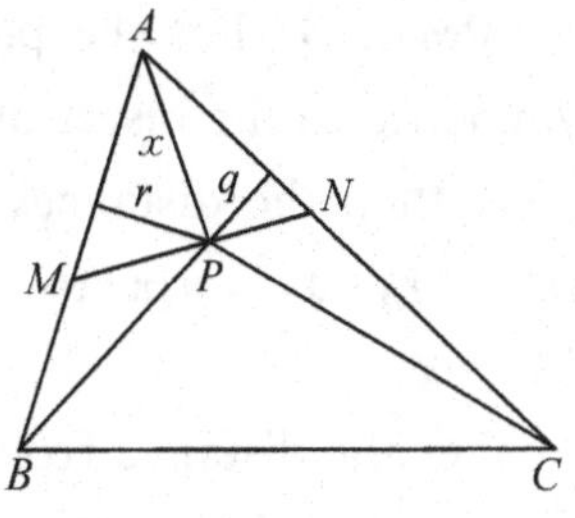

We have

$$\frac{AN}{MN} = \frac{c}{a}, \frac{AM}{MN} = \frac{b}{a}.$$

Figure 5.2

Since

$$S_{\triangle AMN} = S_{\triangle AMP} + S_{\triangle ANP},$$

we have

$$AP \cdot MN \geqslant q \cdot AN + r \cdot AM.$$

So that

$$x = AP \geqslant q \cdot \frac{AN}{MN} + r \cdot \frac{AM}{MN}.$$

Namely

$$x \geqslant \frac{c}{a} \cdot q + \frac{b}{a} \cdot r. \tag{a}$$

Similarly

$$y \geqslant \frac{a}{b} \cdot r + \frac{c}{b} \cdot p, \tag{b}$$

$$z \geqslant \frac{b}{c} \cdot p + \frac{a}{c} \cdot q. \tag{c}$$

Adding up inequalities (a), (b), (c), we get

$$\begin{aligned} x + y + z &\geqslant p\left(\frac{c}{b} + \frac{b}{c}\right) + q\left(\frac{c}{a} + \frac{a}{c}\right) + r\left(\frac{b}{a} + \frac{a}{b}\right) \\ &\geqslant 2(p + q + r). \end{aligned}$$

□

The following method of symmetric point has been noticed by lots of people. Here we adopt Mr. Zou Ming's proof, which is concise and comprehensible.

Proof. 3. Let the point P' and P be symmetric to the bisect of $\angle A$ (see Figure 5.3), then the distances from P' to CA, AB is r, q, respectively, and $P'A = PA = x$.

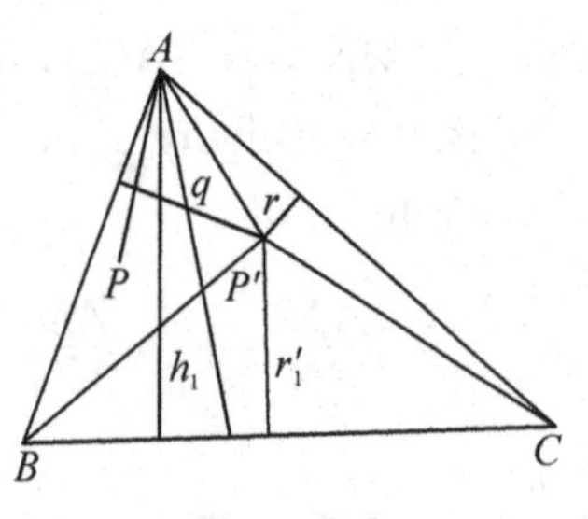

Figure 5.3

Let the distance from A, P' to BC be h_1, r'_1 respectively, then

$$P'A + r'_1 = PA + r'_1 \geqslant h_1,$$

multiply by a on both sides, we have

$$a \cdot PA + ar'_1 \geqslant ah_1$$

$$= 2S_{\triangle ABC}$$
$$= ar'_1 + cq + br.$$

So that

$$x \geqslant \frac{c}{a} \cdot q + \frac{b}{a} \cdot r,$$

similarly

$$y \geqslant \frac{a}{b} \cdot r + \frac{c}{b} \cdot p,$$
$$z \geqslant \frac{b}{c} \cdot p + \frac{a}{c} \cdot q.$$

Adding up above three inequalities, we get

$$\begin{aligned} x + y + z &\geqslant p\left(\frac{c}{b} + \frac{b}{c}\right) + q\left(\frac{c}{a} + \frac{a}{c}\right) + r\left(\frac{b}{a} + \frac{a}{b}\right) \\ &\geqslant 2(p + q + r). \end{aligned}$$

□

The following proof has been noticed much early. The key to the proof is to consider the bisectors in the triangle and applying the embedding inequality.

Proof. 4. (See Figure 5.4.) Denote $\angle BPC = 2\alpha$, $\angle CPA = 2\beta$, $\angle APB = 2\gamma$. Let their bisectors be ω_a, ω_b, ω_c respectively. We only need to prove the following stronger inequality

$$x + y + z \geqslant 2(\omega_a + \omega_b + \omega_c).$$

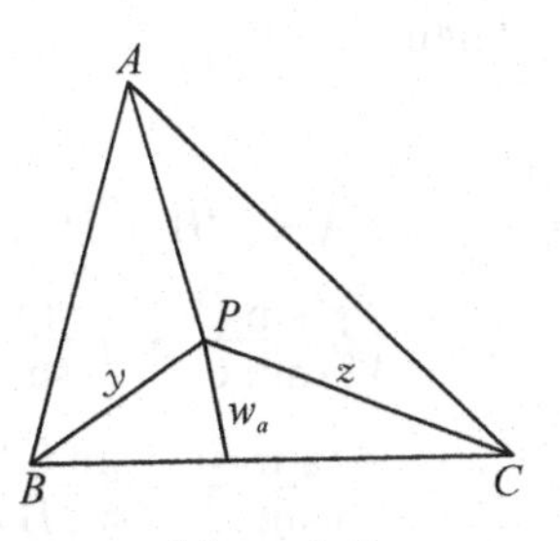

Figure 5.4

By the formula of angle bisector, we obtain

$$\omega_a = \frac{2yz}{y+z} \cos \frac{1}{2} \angle BPC \leqslant \sqrt{yz} \cos \alpha.$$

Similarly

$$\omega_b \leqslant \sqrt{zx} \cos \beta,$$

$$\omega_c \leqslant \sqrt{xy}\cos\gamma.$$

Since $\alpha+\beta+\gamma=\pi$, by the embedding inequality, we conclude that

$$2(\omega_a+\omega_b+\omega_c) \leqslant 2(\sqrt{yz}\cos\alpha+\sqrt{zx}\cos\beta+\sqrt{xy}\cos\gamma)$$
$$\leqslant x+y+z.$$

□

Kang Jiayin, told me the following proof when he was in Grade 2 of Shenzhen High School. He was elected for National Team in 2003.

Proof. 5. (See Figure 5.5.) Make $DT_1 \perp FP$, $ET_2 \perp FP$, the feet are T_1, T_2, respectively.

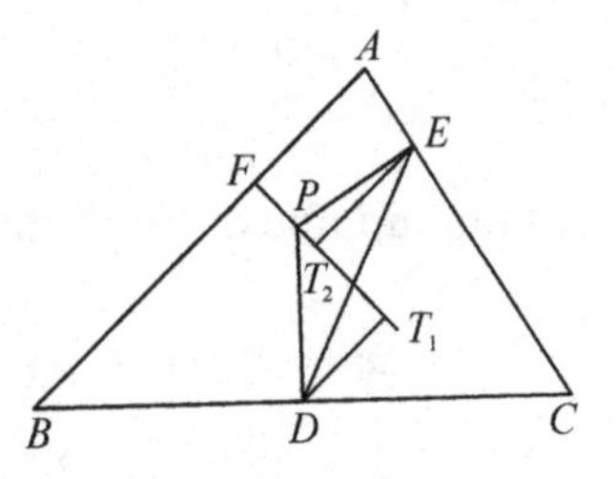

Figure 5.5

Since

$$DE \geqslant DT_1+ET_2,\ DT_1=p\sin B,$$
$$ET_2=q\sin A,$$

we have

$$z=\frac{DE}{\sin C} \geqslant \frac{p\sin B+q\sin A}{\sin C}$$
$$=p\frac{\sin B}{\sin C}+q\frac{\sin A}{\sin C}.$$

Then

$$x+y+z$$
$$=PA+PB+PC$$
$$\geqslant \left(p\frac{\sin B}{\sin C}+q\frac{\sin A}{\sin C}\right)+q\left(\frac{\sin C}{\sin A}+r\frac{\sin B}{\sin A}\right)+\left(r\frac{\sin A}{\sin B}+q\frac{\sin C}{\sin B}\right)$$
$$=p\left(\frac{\sin B}{\sin C}+\frac{\sin C}{\sin B}\right)+q\left(\frac{\sin A}{\sin C}+q\frac{\sin C}{\sin A}\right)+r\left(\frac{\sin B}{\sin A}+q\frac{\sin A}{\sin B}\right)$$
$$\geqslant 2(p+q+r).$$

□

Remark. There have been lots of results about the Erdös-Mordell's inequality. Its generalization on plane is easy, which had been finished early by N. Ozeki and H. Vigler. Later it was rediscovered by others many times. While its generalization in space, especially in n-dimensional space, is difficult. As far as I know, the ideal result has not been obtained.

Example 2. Denote a, b and c the three sides of $\triangle ABC$, then

$$h_a + m_b + t_c \leqslant \frac{\sqrt{3}}{2}(a + b + c), \tag{2}$$

where h_a, m_b and t_c are the altitude of BC, the mid-line of AC and the bisector of $\angle C$ respectively.

Proof. (See Figure 5.6.) Consider the bisector t_a of $\angle A$ instead of altitude h_b. We are to prove a stronger inequality:

$$t_a + m_b + t_c \leqslant \frac{\sqrt{3}}{2}(a + b + c). \tag{a}$$

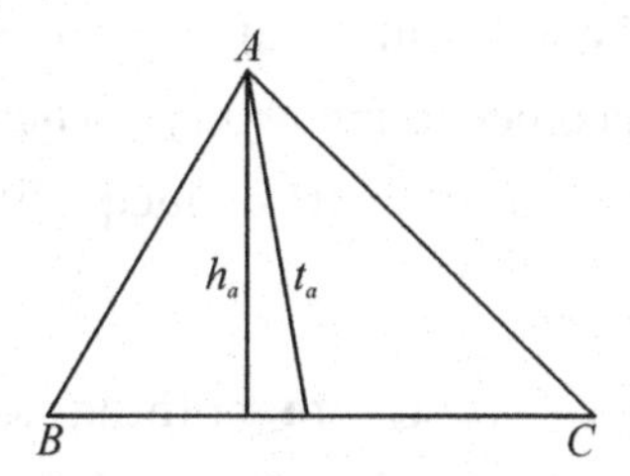

Figure 5.6

In order to prove (a), it suffices to prove a partial inequality

$$m_b + 2t_a \leqslant \frac{\sqrt{3}}{2}(b + 2c). \tag{b}$$

If (b) is true, similarly we have

$$m_b + 2t_c \leqslant \frac{\sqrt{3}}{2}(b + 2a). \tag{c}$$

Adding up (b) and (c), we obtain (a). So we need only to prove (b). By the formula of angle bisector, we have

$$\begin{aligned} t_a^2 &= \frac{4}{(b+c)^2} \cdot bcp(p-a) \\ &\leqslant p(p-a) = \frac{1}{4}((b+c)^2 - a^2). \end{aligned} \tag{d}$$

Notice that

$$m_b^2 = \frac{1}{4}(2a^2 + 2c^2 - b^2). \tag{e}$$

By Cauchy's inequality and (d), (e), we conclude that

$$\begin{aligned} m_b + 2t_a &\leqslant \sqrt{3(m_b^2 + 2t_a^2)} \\ &\leqslant \sqrt{\frac{3}{4}(2a^2 + 2c^2 - b^2 + 2(b+c)^2 - 2a^2)} \\ &= \frac{\sqrt{3}}{2}(b + 2c), \end{aligned}$$

which is (b). □

Remark. (1) Carefully observing various proofs of Examples 1 and 2, we assure that it is a common technique to consider a part of linear geometric inequality. The aim of various proofs of Example 1 is to obtain the part of inequality

$$x \geqslant \lambda_1 q + \lambda_2 r.$$

λ_1, λ_2 are nothing to do with the moving point P. While in Example 2 we obtain the result by find the local inequality

$$m_b + 2t_a \leqslant \frac{\sqrt{3}}{2}(b + 2c).$$

(2) In Example 2, we make stronger proposition by consider the angle bisector instead of the altitude. This useful technique is adopted in Proof 4 to Example 1, which will be adopted again in Example 5 of the last chapter "Tetrahedral Inequality".

Example 3. Given an acute triangle ABC. Denote h_a, h_b, h_c the altitude of sides BC, CA, AB, respectively, and s the semi-circumference. Then

$$\sqrt{3} \cdot \max\{h_a, h_b, h_c\} \geqslant s.$$

Equality holds if $\triangle ABC$ is equilateral.

Proof. If $\triangle ABC$ is not equilateral, the problem can be changed into a problem for the isosceles triangle.

In fact, if $\angle A \geqslant \angle B > \angle C$, then $\angle A > \frac{\pi}{3}$, and $h_c > h_b \geqslant h_a$. Denote h the longest altitude h_c (see Figure 5.7). Extend the shortest side AB to D with $AD = AC$ and link CD. If $\sqrt{3}h \geqslant s$ holds for isosceles $\triangle ACD$, then it holds for general acute triangle.

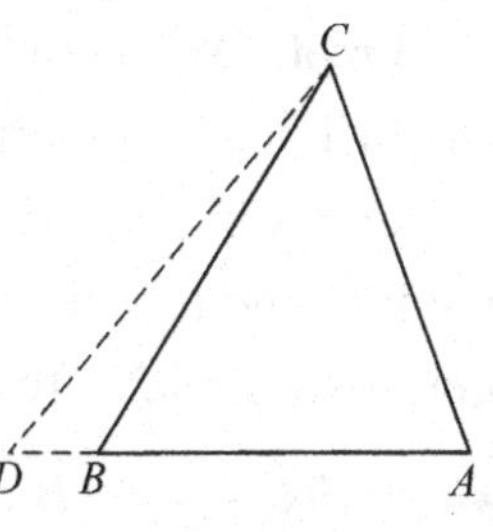

Figure 5.7

We first prove $\sqrt{3}h \geqslant s$ for the isosceles triangle. Since

$$s = AC + \frac{1}{2}CD,\ CD = 2AC \cdot \sin\frac{A}{2},\ h_c = AC \cdot \sin A,$$

we conclude that $\sqrt{3}h \geqslant s$ is equivalent to

$$\sqrt{3}\sin A \geqslant 1 + \sin\frac{A}{2} \left(\frac{\pi}{3} < A < \frac{\pi}{2}\right). \tag{a}$$

Denote $x = \sin\frac{A}{2}$, then $\frac{1}{2} < x < \frac{\sqrt{2}}{2}$, (a) changes into

$$12x^4 - 11x^2 + 2x + 1 \leqslant 0.$$

Namely

$$(2x - 1)(x + 1)(6x^2 - 3x - 1) \leqslant 0. \tag{b}$$

Notice that the range of variable x, it is easy to see $2x - 1 > 0$, $x + 1 > 0$, $6x^2 - 3x - 1 \leqslant 0$, so that (b) follows. □

Remark. The technique of change the general triangle into isosceles triangle is worthy to be noticed. It greatly simplifies the problem.

Example 4 (Zirakzadeh's inequality). Suppose that points P, Q, R lie on three sides BC, CA, AC and trisect the perimeter of $\triangle ABC$, then

$$QR + RP + PQ \geqslant \frac{1}{2}(a + b + c).$$

Proof. We adopt the following projection method to produce a part of linear geometric inequality.

(See Figure 5.8.) Draw two lines from points Q and R perpendicular to line BC, the feet are M and N, respectively, then

$$QR \geqslant MN = a - (BR \cdot \cos B + CQ \cdot \cos C).$$

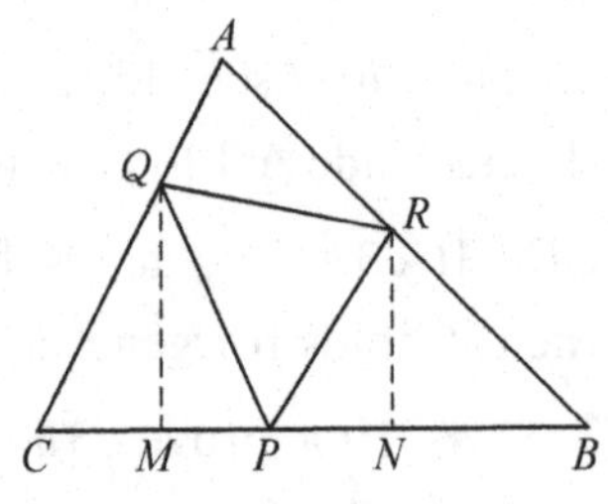

Figure 5.8

Similarly

$$RP \geqslant b - (CP \cdot \cos C + AR \cdot \cos A),$$
$$PQ \geqslant c - (AQ \cdot \cos A + BP \cdot \cos B).$$

Adding up above three inequalities and notice that

$$AQ + AR = BR + BP = CP + CQ = \frac{1}{3}(a + b + c),$$

we have,

$$QR + RP + PQ \geqslant \frac{1}{3}(a + b + c)(3 - \cos A - \cos B - \cos C),$$

by

$$\cos A + \cos B + \cos C \leqslant \frac{3}{2},$$

we conclude that

$$QR + RP + PQ \geqslant \frac{1}{2}(a + b + c).$$

□

Remark. The above beautiful answer was given by Mr. Yang Xuezhi. This problem ever caused extensive discussion.

The following difficult problem of Example 5 was found and proved by Mr. Wang Zhen.

Example 5. Suppose that I, G are the incenter and the barycenter of $\triangle ABC$, respectively, then

$$AI + BI + CI \leqslant AG + BG + CG.$$

Proof. Denote $BC = a$, $AC = b$, $AB = c$. Without loss of generality, we may assume that $a \geqslant b \geqslant c$. (See Figure 5.9.) We will prove that G must lie on $\triangle BIC$.

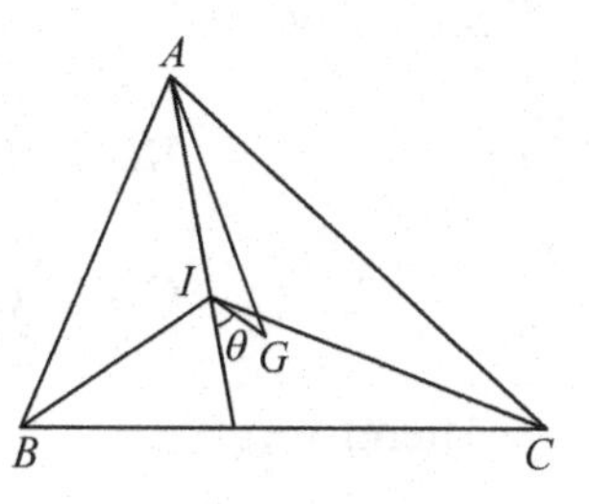

Figure 5.9

Firstly, we will prove that G can not lie in $\triangle AIB$. Otherwise suppose G were in $\triangle AIB$, then

$$S_{\triangle ABG} < S_{\triangle AIB}.$$

Notice that

$$S_{\triangle ABG} = \frac{1}{3} S_{\triangle ABC}, \qquad \frac{S_{\triangle AIB}}{S_{\triangle ABC}} = \frac{c}{a+b+c} \leqslant \frac{1}{3},$$

we obtain

$$S_{\triangle AIB} \leqslant \frac{1}{3} S_{\triangle ABC} = S_{\triangle ABG}.$$

It is a contradiction.

Secondly, we prove G cannot lie in $\triangle AIC$. Otherwise suppose G were in $\triangle AIC$. Suppose that CI intersects AB at T, and CG intersects AB at L, then $AT > AL$.

Notice that

$$AL = BL, \qquad \frac{AT}{BT} = \frac{b}{a} \leqslant 1,$$

therefore

$$AT \leqslant \frac{1}{2} AB = AL.$$

It is a contradiction.

We conclude that G lies on $\triangle BIC$, furthermore G lies on the right side of AI. Let θ be the supplementary angle of $\angle AIG$, then $0 \leqslant \theta \leqslant \frac{A+C}{2}$. We have

$$AG \geqslant AI + GI\cos\theta.$$

Similarly

$$BG \geqslant BI + GI\cos\left(90^\circ + \frac{C}{2} - \theta\right),$$

$$CG \geqslant CI - GI\cos\left(\frac{A+C}{2} - \theta\right).$$

Therefore

$$\begin{aligned} & AG + BG + CG - (AI + BI + CI) \\ \geqslant\ & GI\left(\cos\theta + \cos\left(90^\circ + \frac{C}{2} - \theta\right) - \cos\left(\frac{A+C}{2} - \theta\right)\right) \\ =\ & GI\left(\cos\theta - 2\sin\frac{B+C}{4}\cos\left(\frac{B-C}{4} + \theta\right)\right). \end{aligned}$$

Notice that

$$\frac{B+C}{4} \leqslant 30^\circ,\ \theta \leqslant \frac{B-C}{4} + \theta < 90^\circ,$$

we have

$$\cos\theta - 2\sin\frac{B+C}{4}\cos\left(\frac{B-C}{4} + \theta\right) \geqslant \cos\theta - \cos\left(\frac{B-C}{4} + \theta\right) \geqslant 0,$$

so that

$$AG + BG + CG \geqslant AI + BI + CI.$$

□

Exercises 5

1. Let G be the barycenter of $\triangle ABC$. AG, BG, CG intersect circumcircle of $\triangle ABC$ at points A_1, B_1, C_1, respectively. Then

$$GA_1 + GB_1 + GC_1 \geqslant GA + GB + GC.$$

Equality holds if and only if $\triangle ABC$ is equilateral.

2. For given four points in a convex quadrangle, show that there is a point on boundary of the quadrangle, so that the sum of distance from the point to vertexes of quadrangle is greater than that of the distance from it to four given points. (A problem of St. Petersburg Mathematical Contest in 1993.)

3. Suppose that $ABCDEF$ is a convex hexagon, and $AB \parallel ED$, $BC \parallel FE$, $CD \parallel AF$. Denote R_A, R_C, R_E the radius of circumcircle of $\triangle FAB$, $\triangle BCD$, $\triangle DEF$ respectively, and p is the perimeter of hexagon, show that

$$R_A + R_C + R_E \geqslant \frac{p}{2}.$$

(A problem of 37th IMO.)

4. (Cavachi) Suppose a is the longest side of convex hexagon $ABCDEF$, and $d = \min\{AD, BE, CF\}$, then

$$d \leqslant 2a.$$

5. (Zhu Jiegen) Suppose that I is the incencer of $\triangle ABC$. Denote r_1, r_2, r_3 the radius of inscribed circle of $\triangle IBC$, $\triangle ICA$, $\triangle IAB$, respectively, show that

$$3\sqrt{3}(2-\sqrt{3})r \leqslant r_1 + r_2 + r_3 \leqslant \frac{3\sqrt{3}(2-\sqrt{3})}{2}R,$$

where r, R are the radius of inscribed circle and circumcircle of $\triangle ABC$, respectively.

Chapter 6 Algebraic methods

So far the methods we used are mostly of geometric and triangular. In this section we mainly introduce the algebraic method.

It is convenient to construct algebra identities to prove some distance inequalities. The following typical inequality was given by M. S. Klamkin at his early times.

Example 1. On a plane, there is a $\triangle ABC$ and a point P. Show that

$$a \cdot PB \cdot PC + b \cdot PC \cdot PA + c \cdot PA \cdot PB \geqslant abc.$$

Proof. We consider the plane as a complex plane. Let P, A, B, C correspond to complex numbers z, z_1, z_2, z_3 respectively. Define

$$f(z) = \frac{(z-z_2)(z-z_3)}{(z_1-z_2)(z_1-z_3)} + \frac{(z-z_3)(z-z_1)}{(z_2-z_3)(z_2-z_1)} + \frac{(z-z_1)(z-z_2)}{(z_3-z_1)(z_3-z_2)},$$

then $f(z)$ is a quadratic polynomial of z. Notice that

$$f(z_1) = f(z_2) = f(z_3) = 1,$$

hence $f(z) \equiv 1$. So that

$$\frac{PB \cdot PC}{bc} + \frac{PC \cdot PA}{ca} + \frac{PA \cdot PB}{ab}$$
$$= \left|\frac{(z-z_2)(z-z_3)}{(z_1-z_2)(z_1-z_3)}\right| + \left|\frac{(z-z_3)(z-z_1)}{(z_2-z_3)(z_2-z_1)}\right| + \left|\frac{(z-z_1)(z-z_2)}{(z_3-z_1)(z_3-z_2)}\right|$$
$$\geqslant |f(z)| = 1.$$

□

The following interesting problem was proposed by Tweedie.

Example 2. Suppose $\triangle ABC$, $\triangle A'B'C'$ are equilateral triangles on a plane with the same direction of vertex array, then the sum of any two of line segments AA', BB', CC' is greater than or equal to the third one.

Proof. (See Figure 6.1.) Since $\triangle ABC$, $\triangle A'B'C'$ are similar and with the same direction of vertex array, then

$$(z_1' - z_1)(z_2 - z_3) + (z_2' - z_2)(z_3 - z_1) + (z_3' - z_3)(z_1 - z_2) = 0,$$

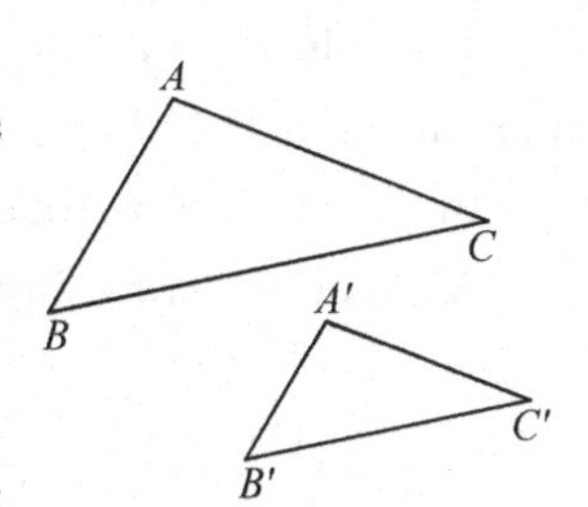

Figure 6.1

where A, B, C correspond to complex numbers z_1, z_2, z_3; A', B', C' correspond to complex numbers z_1', z_2', z_3'. By the property of complex norm, we get

$$|z_1' - z_1| \cdot |z_2 - z_3| + |z_2' - z_2| \cdot |z_3 - z_1| \geqslant |(z_3' - z_3)(z_1 - z_2)|.$$

Notice that $\triangle ABC$ is equilateral, so that

$$|z_2 - z_3| = |z_3 - z_1| = |z_1 - z_2|.$$

Therefore

$$|z_1' - z_1| + |z_2' - z_2| \geqslant |z_3' - z_3|.$$

Namely

$$AA' + BB' \geqslant CC'.$$

Similarly we can get the other two inequalities. □

Now we recall a simple proposition in plane geometry: three positive numbers a, b, c can be three sides of a triangle if and only if there exists three positive numbers x, y, z such that $a = y + z$, $b = x + z$, $c = x + y$. (The sufficiency of this conclusion can be verified directly. Decomposition of

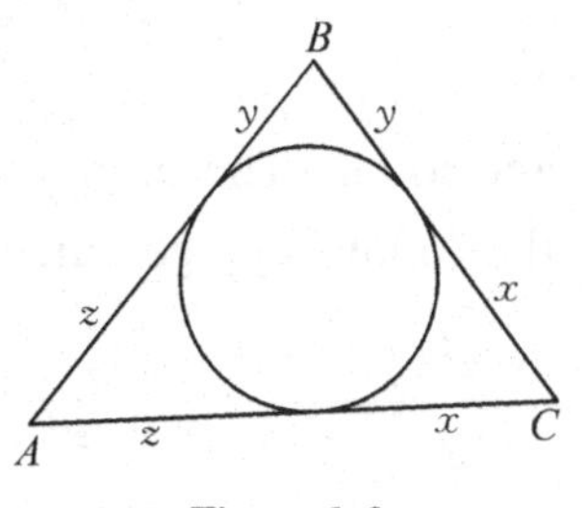

Figure 6.2

Figure 6.2 shows the necessity.)

By this conclusion, we can consider the inequality of x, y, z instead of inequality of triangle sides a, b, c by the identities $x = -a+b+c$, $y = a-b-c$, $z = a+b-c$.

The solution of following question in the 24th IMO is a typical application of the above method.

In $\triangle ABC$, show that $b^2c(b-c)+c^2a(c-a)+a^2b(a-b)\geqslant 0$.

A concise proof using the above method is to use the inequality:

$$\left(\frac{x^2}{y}+\frac{y^2}{z}+\frac{z^2}{x}\right)(y+z+x)\geqslant (x+y+z)^2,$$

by Cauchy inequality. The detail answer can be seen in any IMO tutorial book.

The following problem is a bit new.

Example 3. Denote r_a, r_b, r_c the radius of escribed circles corresponding to three sides a, b, c of $\triangle ABC$ respectively, show that

$$\frac{a^2}{r_b^2+r_c^2}+\frac{b^2}{r_c^2+r_a^2}+\frac{c^2}{r_a^2+r_b^2}\geqslant 2.$$

Proof. By the substitution

$$x=-a+b+c,\quad y=a-b+c,\quad z=a+b-c,$$

so that x, y, $z>0$. Notice that

$$S_{\triangle ABC}=\frac{1}{4}\sqrt{(x+y+z)xyz},$$

$$r_a=\frac{2S_{\triangle ABC}}{b+c-a}=\frac{1}{2x}\sqrt{(x+y+z)xyz},$$

and so on. The original inequality is equivalent to (after calculating) the following algebraic inequality.

$$\frac{y^2z^2(y+z)^2}{y^2+z^2}+\frac{z^2x^2(z+x)^2}{z^2+x^2}+\frac{x^2y^2(x+y)^2}{x^2+y^2}\geqslant 2xyz(x+y+z). \tag{a}$$

To prove (a), it suffices to prove the following interesting partial inequality

$$\frac{y^2z^2(y+z)^2}{y^2+z^2} \geqslant \frac{2xyz(x+y+z)y^2z^2}{x^2y^2+y^2z^2+z^2x^2}. \tag{b}$$

In fact, if (b) is established, we add this kind of three inequalities together, then the desired result follows.

To prove (b) as follows.

$$\begin{aligned}
(2) &\Leftrightarrow (y+z)^2(x^2y^2+y^2z^2+z^2x^2) \geqslant 2xyz(x+y+z)(y^2+z^2),\\
&\Leftrightarrow (y^2+z^2)x^2(y+z)^2+y^2z^2(y+z)^2 \geqslant 2xyz(x+y+z)(y^2+z^2),\\
&\Leftrightarrow (y^2+z^2)^2x^2+y^2z^2(y+z)^2 \geqslant 2xyz(y+z)(y^2+z^2),\\
&\Leftrightarrow [(y^2+z^2)x-yz(y+z)]^2 \geqslant 0.
\end{aligned}$$

□

Remark. The proof to algebraic inequality (a) is very difficult. Mr. Lin Song told me the above method of changing into partial inequality (b). (He was then a student of the high school affiliate to Huanan Normal University, and was selected to National Training Team in 2003.)

We will talk about coordinate method in dealing with geometric inequality as follows.

Example 4. Denote $P_i(x_i, y_i)$ $(i=1, 2, 3;\ x_1<x_2<x_3)$ the points on rectangular coordinate plane, and R the radius of circumscribed circle of $\triangle P_1P_2P_3$, then

$$\frac{1}{R}<2\left|\frac{y_1}{(x_1-x_2)(x_1-x_3)}+\frac{y_2}{(x_2-x_1)(x_2-x_3)}+\frac{y_3}{(x_3-x_1)(x_3-x_2)}\right|,$$

show that coefficient 2 is the best possible.

Proof. Since the shift of $\triangle P_1P_2P_3$ along x-axis, x_1-x_2, x_2-x_3, x_3-x_1 are unchanged, the value on both sides of the original inequality are unchanged.

When we shift $\triangle P_1P_2P_3$ along y-axis, notice that

$$\frac{1}{(x_1 - x_2)(x_1 - x_3)} + \frac{1}{(x_2 - x_1)(x_2 - x_3)} + \frac{1}{(x_3 - x_1)(x_3 - x_2)} = 0.$$

Thus the value on both sides of the original inequality are unchanging.

Therefore, we may assume P_1 be the origin (x_1, $x_2 = 0$), so that the original inequality is changed into

$$\frac{1}{2R} < \frac{1}{|x_2 - x_3|}\left|\frac{y_3}{x_3} - \frac{y_2}{x_2}\right|.$$

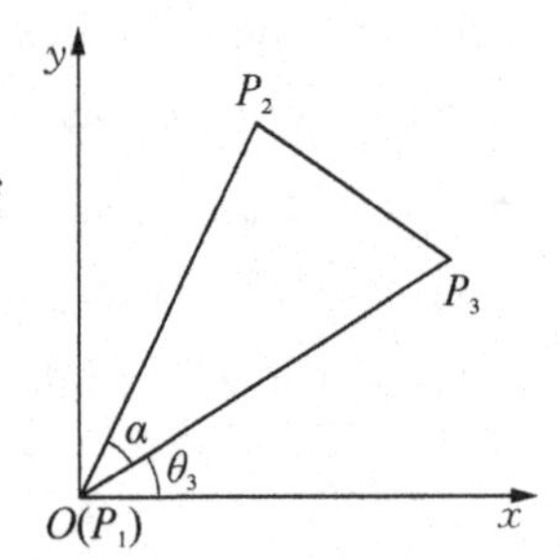

Figure 6.3

(See Figure 6.3.) Let θ_3, θ_2 be angles of inclination of lines OP_3, OP_2. $\angle P_2P_1P_3 = \alpha$, then $\theta_3 - \theta_2 = \pm\alpha$, therefore,

$$\begin{aligned}
&\frac{1}{|x_2 - x_3|}\left|\frac{y_3}{x_3} - \frac{y_2}{x_2}\right| \\
&= \frac{1}{|x_2 - x_3|}|\tan\theta_3 - \tan\theta_2| \\
&= \frac{1}{|x_2 - x_3|}\left|\frac{\sin(\theta_3 - \theta_2)}{\cos\theta_3\cos\theta_2}\right| \\
&= \frac{\sin\alpha}{|x_2 - x_3|}\left|\frac{1}{\cos\theta_3\cos\theta_2}\right| \\
&> \frac{\sin\alpha}{|x_2 - x_3|} \geqslant \frac{\sin\alpha}{P_2P_3} = \frac{1}{2R}.
\end{aligned}$$

If $y_1 = y_3 = x_1 = 0$, and $y_2 \to 0$, then $\cos\theta_3 = 1$, $\cos\theta_2 \to 1$. The left side of above inequality $\to \frac{1}{2R}$, hence 2 is the best possible. □

Remark. The above solution was given by student Xiang Zhen.

Example 5. Let P be any point in the plane of acute $\triangle ABC$. Denote u, v, w the distances from P to A, B, C, respectively. Prove that

$$u^2\tan A + v^2\tan B + w^2\tan C \geqslant 4\Delta,$$

and give the condition such that equality holds, where Δ denotes the area of $\triangle ABC$.

Proof. We establish planar Cartesian coordinate system by taking the line BC for x-axis and the line of the altitude passing A for y-axis (see Figure 6.4).

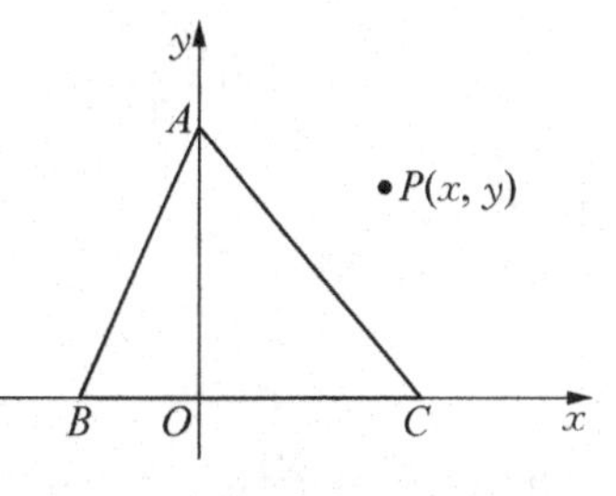

Figure 6.4

The coordinates of A, B, C are $(0, a)$, $(-b, 0)$, $(c, 0)$ (where a, b, $c > 0$), then

$$\tan B = \frac{a}{b}, \qquad \tan C = \frac{a}{c},$$

$$\tan A = -\tan(B + C) = \frac{a(b + c)}{a^2 - bc}.$$

By $\angle A$ is acute angle, we have $a^2 - bc > 0$.

Suppose the coordinate of P is (x, y), then

$$\begin{aligned}
& u^2 \tan A + v^2 \tan B + w^2 \tan C \\
= & [x^2 + (y - a)^2] \frac{a(b + c)}{a^2 - bc} + \frac{a}{b}[(x + b)^2 + y^2] + \frac{a}{c}[(x - c)^2 + y^2] \\
= & (x^2 + y^2 + a^2 - 2ay) \frac{a(b + c)}{a^2 - bc} + \frac{a(b + c)}{bc}(x^2 + y^2 + bc) \\
= & \frac{a(b + c)}{bc(a^2 - bc)}[a^2 x^2 + (ay - bc)^2 + 2bc(a^2 - bc)] \\
\geqslant & \frac{a(b + c)}{bc(a^2 - bc)} \cdot 2bc(a^2 - bc) \\
= & 2a(b + c) = 4\Delta.
\end{aligned}$$

From above proof, we see that equality holds if and only if $x = 0$ and $y = bc/a$, namely, P is the orthocentre $(0, bc/a)$ of $\triangle ABC$. □

The discovery and construction of algebraic identities is the most basic method to find and prove geometric inequalities.

Example 6. Let a, b, c be three sides of $\triangle ABC$, respectively. a', b', c' are three sides of $\triangle A'B'C'$. $S_{\triangle ABC} = F$ and $S_{\triangle A'B'C'} = F'$. Suppose that

$$\mu = \min\left\{\frac{a^2}{a'^2}, \frac{b^2}{b'^2}, \frac{c^2}{c'^2}\right\}, \quad \nu = \max\left\{\frac{a^2}{a'^2}, \frac{b^2}{b'^2}, \frac{c^2}{c'^2}\right\},$$

then

$$H \geqslant 8\left(\lambda F'^2 + \frac{1}{\lambda}F^2\right),$$

where $H = a'^2(-a^2 + b^2 + c^2) + b'^2(a^2 - b^2 + c^2) + c'^2(a^2 + b^2 - c^2)$.

Proof. By Heron's formula for triangle area, we have

$$16F^2 = (a^2 + b^2 + c^2)^2 - 2(a^4 + b^4 + c^4),$$

$$16F'^2 = (a'^2 + b'^2 + c'^2)^2 - 2(a'^4 + b'^4 + c'^4).$$

Denote $D_1 = \sqrt{\lambda}a'^2 - \frac{a^2}{\sqrt{\lambda}}$, $D_2 = \sqrt{\lambda}b'^2 - \frac{b^2}{\sqrt{\lambda}}$, $D_3 = \sqrt{\lambda}c'^2 - \frac{c^2}{\sqrt{\lambda}}$, then we have the identity:

$$H - 8\left(\lambda F'^2 + \frac{1}{\lambda}F^2\right) = \frac{1}{2}D_1^2 - D_1(D_2 + D_3) + \frac{1}{2}(D_2 - D_3)^2. \tag{a}$$

Notice that when $\lambda = \mu = \frac{a^2}{a'^2}$, it follows that $D_1 = 0$, therefore by (a), we have

$$H - 8\left(\frac{a^2}{a'^2}F'^2 + \frac{a'^2}{a^2}\right)F^2 = \frac{1}{2}\left[\frac{a}{a'}(b'^2 - c'^2) - \frac{a'}{a}(b^2 - c^2)\right]^2 \geqslant 0,$$

namely,

$$H \geqslant 8\left(\mu F'^2 + \frac{1}{\mu}F^2\right). \tag{b}$$

Similarly

$$H \geqslant 8\left(\nu F'^2 + \frac{1}{\nu}F^2\right). \tag{c}$$

For given $\lambda \in [\mu, \nu]$, let

$$\lambda = \theta\mu + (1 - \theta)\nu, \; 0 \leqslant \theta \leqslant 1.$$

Let (b), (c) be multiplied by θ and $1-\theta$, respectively, then adding them up together, we have

$$H \geqslant 8\left[\lambda F'^2 + \left(\frac{\theta}{\mu} + \frac{1-\theta}{\nu}\right)F^2\right]. \tag{d}$$

It is easy to see

$$\frac{\theta}{\mu} + \frac{1-\theta}{\nu} \geqslant \frac{1}{\lambda}.$$

Then by (d), it follows that

$$H \geqslant 8\left(\lambda F'^2 + \frac{1}{\lambda}F^2\right).$$

□

Remark. It is the key difficulty to find algebraic identity (a) in above example. This example given by Mr. Chen Ji is a strengthened form to Neuberg-Pedoe's inequality.

Exercises 6

1. Let p be any point in acute $\triangle ABC$, then

$$PA \cdot PB \cdot AB + PB \cdot PC \cdot BC + PC \cdot PA \cdot CA \geqslant AB \cdot BC \cdot CA.$$

Equality holds if and only if P is the orthocenter of $\triangle ABC$.

2. To make two squares $ABDE$ and $ACFG$ outward with sides AB and CD of $\triangle ABC$, respectively. $BP \perp BC$, $CQ \perp BC$, the feet are P, Q, then

$$BP + CQ \geqslant BC + EG.$$

Equality holds if and only if $AB = AC$.

3. Let $P(z)$ be correspond to complex number z on the complex plane. Complex number $a = p + \mathrm{i}q$ (p, $q \in \mathbf{R}$). $P(z_1)$, ..., $P(z_5)$ are the vertexes of convex pentagon Q. Furthermore, the origin and $P(az_1)$, ..., $P(az_5)$ lie in the interior of Q. Prove that

$$p + q \cdot \tan\frac{\pi}{5} \leqslant 1.$$

4. Suppose that a, b, c are three sides of $\triangle ABC$, and h_a, h_b, h_c are altitudes corresponding to a, b, c, respectively, and r_a, r_b, r_c are radius of escribed circle corresponding to a, b, c, respectively. Prove that

$$\left(\frac{h_a}{r_b}\right)^2 + \left(\frac{h_b}{r_c}\right)^2 + \left(\frac{h_c}{r_a}\right)^2 \geqslant 4\left(\sin^2\frac{A}{2} + \sin^2\frac{B}{2} + \sin^2\frac{C}{2}\right).$$

Equality holds if and only if $\triangle ABC$ is equilateral.

5. (Wen Jiajin) Suppose that AD, BE, CF are the angle bisectors of $\triangle ABC$. The square roots of the distances from the moving point P within the $\triangle ABC$ to the three sides are the lengths of sides of a triangle. Prove that.

(1) The orbit of P is in the interior of a ellipse Γ. And Γ is tangent to the three sides BC, AB, AC of $\triangle ABC$ at points D, E, F.

(2) The area S_Γ of Γ satisfies

$$\frac{4\sqrt{3}\pi}{9} S_{\triangle DEF} \geqslant S_\Gamma \geqslant \frac{\sqrt{3}\pi}{9} S_{\triangle ABC}.$$

Chapter 7 Isoperimetric and extremal value problem

All kinds of isoperimetric problems in any space seem to be one of the permanent subjects in geometry studying. Isoperimetric theorem indicates that several special geometric graph in plane geometry, such as circle, regular n-gon. That special properties of them are very spectacular. There are several isoperimetric theorem which are usually used in high school mathematic contents.

Theorem 1 (Isoperimetric Theorem 1). Of all plane figures with given circumference, the circle has the largest area. And of all plane figures with given area, the circle has the least circumference.

Theorem 2 (Isoperimetric Theorem 2). Of all plane n-gons with given circumference, the regular n-gon has the largest area. And of all plane n-gons with given area, the regular n-gon has the least circumference.

Let a_1, a_2, ..., a_n be the side length of a n-gon with area F, then the following isoperimetric inequality is inferred from Isoperimetric Theorem

$$\left(\sum_{i=1}^{n} a_i\right)^2 \geqslant 4n\tan\frac{\pi}{n} \cdot F.$$

Equality holds if and only if $a_1 = a_2 = \cdots = a_n$.

The following Theorem 3 is a generalization of extremal value properties of inscribed quadrilateral in a circle in Chapter 3.

Theorem 3 (Steiner's Theorem). For all of n-gons with given sides, the one with a circumcircle has the largest area.

Theorem 4. Of all inscribed n-polygon in a circle, the regular n-gon has the largest area.

Let a_1, a_2, $\dots$, a_n be lengths of sides of inscribed n-gon in a circle which radius R, then by Theorem 4, we can infer that

$$\sum_{i=1}^{n} a_i \leqslant 2nR \cdot \sin\frac{\pi}{n},$$

equality holds if and only if $a_1 = a_2 = \cdots = a_n$.

Firstly, we discuss several simple examples.

Example 1. Let r be the radius of a circle. l be the tangent line of the circle through a given point P. Through a moving point R in the circle, make a line RQ perpendicular to l with intersection point Q. Try to find the maximum area of $\triangle PQR$. (The 13th Canada's mathematic problem.)

Answer. (See Figure 7.1.)

Notice that $OP \parallel RQ$, make a line $RS \parallel l$, S is the intersection point with $\odot O$. Join with PS, it is easy to prove that

$$S_{\triangle PQR} = \frac{1}{2} S_{\triangle PRS}.$$

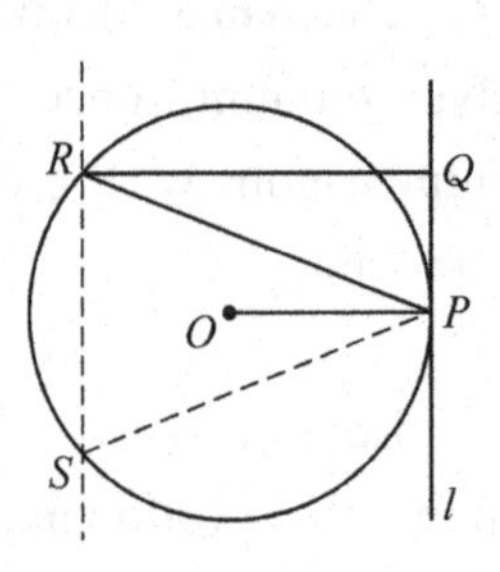

Figure 7.1

By Theorem 4, if inscribed triangle $\triangle PRS$ in a circle is regular, then $\max\{S_{\triangle PRS}\} = \frac{3\sqrt{3}}{4} r^2$, so that $S_{\triangle PQR} = \frac{3\sqrt{3} r^2}{8}$.

The next Example is a familiar problem.

Example 2. (See Figure 7.2.) The curve L divides regular triangle ABC into two sections with equal-area. Show that

$$l \geqslant \frac{\sqrt{\pi}a}{2\sqrt[4]{3}},$$

where l is the length of L, a is the length of side of $\triangle ABC$.

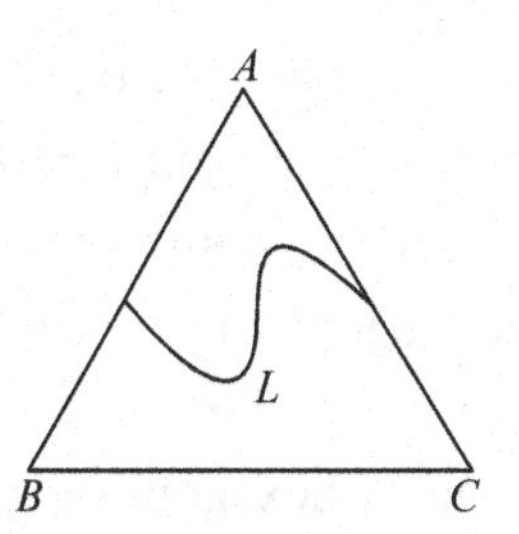

Figure 7.2

Proof. We reflect $\triangle ABC$ on side of $\triangle ABC$ for point A fixed five times (see Figure 7.3), then six L' s form a closed curve. Because the area of the section enclosed by L is a given value $3S_{\triangle ABC}$, from Theorem 1, we can see if the section of closed curve is a circle then it has the minimal circumference.

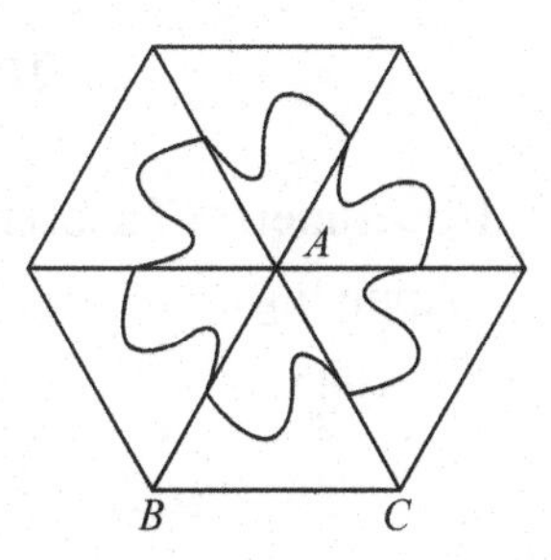

Figure 7.3

So

$$6l \geqslant 2\pi\sqrt{\frac{3S_{\triangle ABC}}{\pi}} = 2\sqrt{\pi} \cdot \sqrt{3 \cdot \frac{\sqrt{3}}{4}a^2}.$$

Hence

$$l \geqslant \frac{\sqrt{\pi}a}{2\sqrt[4]{3}}.$$

□

We discuss Popa's inequality again as follows. (Refer to Example 3 in Chapter 3.)

Example 3. Let Q be a convex quadrilateral with area F and four sides satisfy $a \leqslant b \leqslant c \leqslant d$. Show that

$$F \leqslant \frac{3\sqrt{3}}{4}c^2.$$

The proof using Isoperimetric Theorem is as follows.

Proof. Reflect Q on the longest side to form a hexagon (see Figure 7.4).

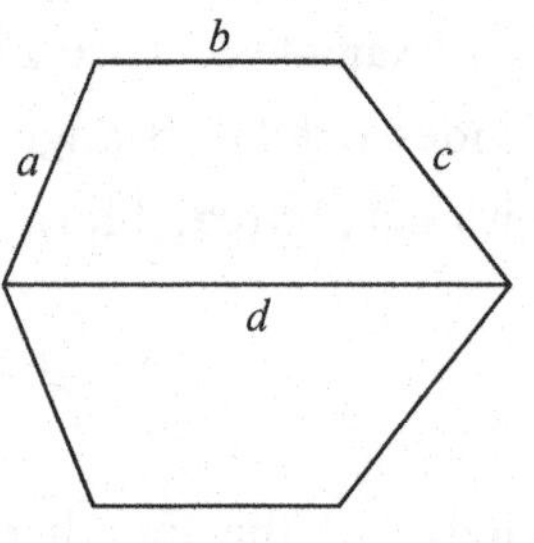

Figure 7.4

(In some special cases will form pentagon or rectangle, but the proof is similar.) Since the circumference of this convex hexagon is a given value $2(a+b+c)$, the area $2F$ of the hexagon satisfy

$$2F \leqslant \frac{3\sqrt{3}}{2}\left(\frac{a+b+c}{3}\right)^2,$$

by Isoperimetric Theorem 2.

Applying a, $b \leqslant c$, we obtain

$$F \leqslant \frac{3\sqrt{3}}{4}c^2.$$

□

Remark. From the above method, we can generalize Popa's inequality to n-gon. (Refer to Problem 5 under Exercises in this chapter.)

By the above method, Ms. Zhang Haijuan and the author construct an inequality about two n-gons as follows.

Example 4. Let Ω_1, Ω_2 be two n-gons with areas F_1, F_2 and sides $a_1 \leqslant a_2 \leqslant \cdots \leqslant a_{n-1} \leqslant a_n$ and $b_1 \leqslant b_2 \leqslant \cdots \leqslant b_{n-1} \leqslant b_n$, respectively. Show that

$$\frac{F_1}{a_n^2} + \frac{F_2}{b_n^2} < \frac{(n-1)^2}{4\pi}\left(\frac{a_{n-1}}{a_n} + \frac{b_{n-1}}{b_n}\right)^2.$$

Proof. Let A_nA_1 be the longest side of Ω_1. Taking A_nA_1 as one side, construct a polygon $\Omega'_2 \sim \Omega_1$ and Ω'_2, Ω_1 locates on different side of A_nA_1. The longest side $B'_nB'_1$ of Ω'_2 superposes A_nA_1. (See Figure 7.5.)

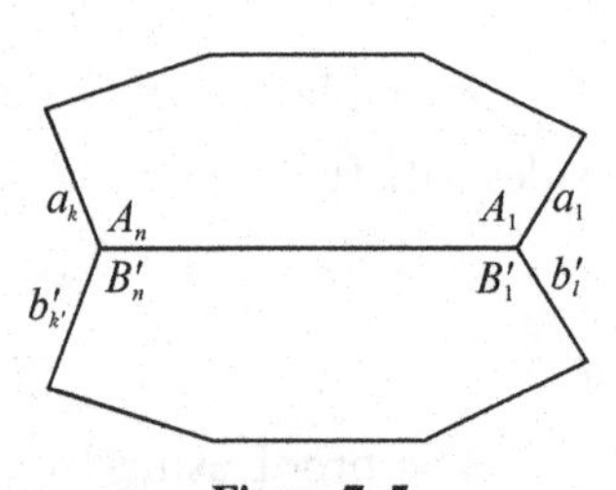

Figure 7.5

Let $\widetilde{\Omega} = \Omega_1 \cup \Omega_2$, then the circumference of $\widetilde{\Omega}$ is

$$\sum_{i=1}^{n-1}\left(a_i + \frac{a_n}{b_n}b_i\right)$$

and the area is $F_1 + \frac{a_n^2}{b_n^2}F_2$.

Let $\angle A_n$, $\angle A_1$ and $\angle B_n$, $\angle B_1$ be two angles of Ω_1, Ω_2, respectively, while the longest side A_nA_1, B_nB_1 belongs to Ω_1, Ω_2, respectively. Let another sides of $\angle A_n$, $\angle A_1$ be a_k, a_1 respectively and also another sides of $\angle B_n$, $\angle B_1$ are $b'_{k'}$, $b'_{l'}$ respectively.

If $\angle A_n$, $\angle A_1$ and $\angle B_n$, $\angle B_1$ are not supplementary angle, respectively, then the sides a_k, a_1 and $b'_{k'}$, $b'_{l'}$ are not in a line, respectively. Then $\widetilde{\Omega}$ is a $2(n-1)$-gon. Applying isoperimetric inequality (Theorem 2), we obtain

$$\begin{aligned} F_1 + \frac{a_n^2}{b_n^2}F_2 &\leqslant \frac{\left(\sum_{i=1}^{n-1}\left(a_i + \frac{a_n}{b_n}b_i\right)\right)^2}{8(n-1)} \cdot \cot\frac{\pi}{2(n-1)} \\ &\leqslant \frac{\left((n-1)a_{n-1} + (n-1)\frac{a_n}{b_n}b_{n-1}\right)^2}{8(n-1)} \cdot \cot\frac{\pi}{2(n-1)} \\ &= \frac{(n-1)\left(a_{n-1} + \frac{a_n}{b_n}b_{n-1}\right)^2}{8} \cdot \cot\frac{\pi}{2(n-1)}, \end{aligned}$$

namely,

$$\frac{F_1}{a_n^2} + \frac{F_2}{b_n^2} \leqslant \frac{(n-1)}{8}\left(\frac{a_{n-1}}{a_n} + \frac{b_{n-1}}{b_n}\right)^2 \cdot \cot\frac{\pi}{2(n-1)}. \tag{a}$$

If the supplementary angle of $\angle A_n$ or $\angle A_1$ is $\angle B_n$ or $\angle B_1$, then $\widetilde{\Omega}$ is a $(2n-3)$-gon. If the supplementary angle of $\angle A_n$, $\angle A_1$ are $\angle B_n$, $\angle B_1$, then $\widetilde{\Omega}$ is a $2(n-2)$-gon. With the same discussion as above, applying isoperimetric inequality, we obtain the following results

$$\frac{F_1}{a_n^2} + \frac{F_2}{b_n^2} \leqslant \frac{(n-1)^2}{4(2n-3)}\left(\frac{a_{n-1}}{a_n} + \frac{b_{n-1}}{b_n}\right)^2 \cdot \cot\frac{\pi}{2n-3}, \tag{b}$$

$$\frac{F_1}{a_n^2}+\frac{F_2}{b_n^2}\leqslant\frac{(n-1)^2}{8(n-2)}\left(\frac{a_{n-1}}{a_n}+\frac{b_{n-1}}{b_n}\right)^2\cdot\cot\frac{\pi}{2(n-2)}. \qquad \text{(c)}$$

Notice that $x\in\left(0,\ \frac{\pi}{2}\right)$, we have

$$\cot x<\frac{1}{x}. \qquad \text{(d)}$$

Now put (a), (b), (c) with (d) together respectively, we get

$$\frac{F_1}{a_n^2}+\frac{F_2}{b_n^2}<\frac{(n-1)^2}{4\pi}\left(\frac{a_{n-1}}{a_n}+\frac{b_{n-1}}{b_n}\right)^2.$$

□

At last, we will introduce a proof to Ozeki's inequality by Isoperimetric Theorem.

Example 5 (Ozeki's inequality). Let $\varphi_1+\varphi_2+\cdots+\varphi_n=\pi(n\geqslant 3)$, $0<\varphi_i<\pi$, $i=1, 2, \ldots, n$. Show that: for every non-negative real numbers $x_1, x_2, \ldots, x_n$, we have

$$\sum_{i=1}^{n}x_i^2\geqslant\sec\frac{\pi}{n}\left(\sum_{i=1}^{n}x_ix_{i+1}\cos\varphi_i\right),$$

where $x_{n+1}=x_1$.

Proof. (See Figure 7.6.) Start with point O and make n rays $OA_1, OA_2, \ldots, OA_n$ satisfying

$$\angle A_iOA_{i+1}=\varphi_i+\frac{\pi}{n},$$

where $A_{n+1}=A_1$. In these n rays, make n segments $OA_i=x_i (i=1, 2, \ldots, n)$, then we get a n-gon $A_1A_2\cdots A_n$.

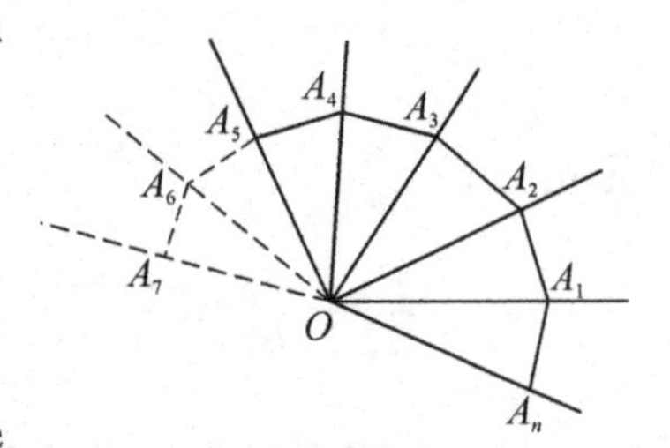

Figure 7.6

Let the area of this n-gon be F. In $\triangle OA_iA_{i+1}$, by cosine law we obtain

$$a_i^2=x_i^2+x_{i+1}^2-2x_i^2x_{i+1}^2\cos\angle A_iOA_{i+1},\ i=1, 2, \ldots, n,\ x_{i+1}=x_1.$$

Hence

$$\sum_{i=1}^{n} a_i^2 = 2\sum_{i=1}^{n} x_i^2 - 2\sum_{i=1}^{n} x_i x_{i+1}\cos\left(\varphi_i + \frac{\pi}{n}\right). \tag{a}$$

By the Cauchy inequality and isoperimetric inequality, we have

$$\sum_{i=1}^{n} a_i^2 \geqslant \frac{1}{n}\left(\sum_{i=1}^{n} a_i\right)^2 \geqslant 4F\tan\frac{\pi}{n}. \tag{b}$$

By (a) and (b), we conclude

$$\begin{aligned}
\sum_{i=1}^{n} x_i^2 &\geqslant 2F\tan\frac{\pi}{n} + \sum_{i=1}^{n} x_i x_{i+1}\cos\left(\varphi_i + \frac{\pi}{n}\right) \\
&= \sum_{i=1}^{n}\left[x_i x_{i+1}\sin\left(\varphi_i + \frac{\pi}{n}\right)\tan\frac{\pi}{n} + x_i x_{i+1}\cos\left(\varphi_i + \frac{\pi}{n}\right)\right] \\
&= \sec\frac{\pi}{n}\sum_{i=1}^{n} x_i x_{i+1}\left[\sin\left(\varphi_i + \frac{\pi}{n}\right)\sin\frac{\pi}{n} + \cos\frac{\pi}{n}\cos\left(\varphi_i + \frac{\pi}{n}\right)\right] \\
&= \sec\frac{\pi}{n}\sum_{i=1}^{n} x_i x_{i+1}\cos\varphi_i.
\end{aligned}$$

□

Remark. (1) Ozeki's inequality was proposed by N. Ozeki when he generalized the famous Erdös-Mordell inequality to polygon. The related results about it could refer to N. Ozeki "On the P. Erdös inequality for the triangle", J. College Arts Sci. Chiba Univ, 1957 (2):247-250. This inequality was recovered by Lenhard in 1961.

(2) Ozeki's inequality is a generalization of triangle imbedding inequality to polygon. It can be generalized to 3 and n-dim spaces. The results for tetrahedron are as follows:

Let θ_{ij} $(1 \leqslant i < j \leqslant 4)$ be inside interfacial angle of tetrahedron Ω, then for each real number $x_1, \ldots, x_4$, we have

$$\sum_{i=1}^{4} x_i^2 \geqslant 2\sum_{1\leqslant i<j\leqslant 4} x_i x_j \cos\theta_{ij}.$$

Other generalizations and related results could refer to the articles of Mr. Zhang Yao and the author (Linear Algebra and its Applications,

1998;278).

(3) Using the above methods in the proof, we can prove the following algebraic inequality:

Suppose that $x, y, z \geqslant 0$, then

$$(x^2+xy+y^2)(y^2+yz+z^2)(z^2+zx+x^2) \geqslant (xy+yz+zx)^2.$$

The proof is left to readers.

Exercises 7

1. Among all of triangles with given side BC and its opposite angle α. Show that:

(1) The isosceles triangle with base BC has the largest area.

(2) The isosceles triangle with base BC has the largest circumference.

2. Two equilateral triangles inscribe a circle of radius r. Let K be the area of overlap section of two triangles. Show that

$$2K \geqslant r^2\sqrt{3}.$$

3. Of all quadrilaterals which three sides each with length 1 and one angle 30°, please find a quadrilateral with the largest area.

4. Let P be a inner point of convex n-gon $A_1A_2\cdots A_n$ and the distance from P to each side $A_1A_2, A_2A_3, \ldots, A_nA_1$ are $d_1, d_2, \ldots, d_n$, respectively. Show that

$$\sum_{i=1}^{n} \frac{a_i}{d_i} \geqslant 2n\tan\frac{\pi}{n},$$

where $a_i = A_iA_{i+1}(A_{n+1} = A_1)$, and give the necessary and sufficient condition for equalities hold.

5. Let a convex n-gon in a plane with area F satisfy $a_1 \leqslant a_2 \leqslant \cdots \leqslant a_n$, show that

$$F < \frac{(n-1)^2}{2\pi}a_{n-1}^2.$$

Chapter 8 Embed inequality and inequality for moment of inertia

Embed triangle inequality (called Embed inequality in short) plays an important role in research of primary geometric inequalities in recent years, which is a source of generating geometry inequalities. Embed triangle inequality can be read as follows.

Theorem 1 (Embedding Triangle Inequality). Suppose that $A + B + C = (2k + 1)\pi$, x, y, $z \in \mathbf{R}$, then

$$x^2 + y^2 + z^2 \geqslant 2yz\cos A + 2zx\cos B + 2xy\cos C, \tag{1}$$

equality holds if and only if $x : y : z = \sin A : \sin B : \sin C$.

Proof. The difference of two sides of (1) is $(x - y\cos C - z\cos B)^2 + (y\sin C - z\sin B) \geqslant 0$. □

By the facial meaning, the geometric explanation that inequality (1) called is: if $0 < A, B, C < \pi$, and for arbitrarily real number x, y, z satisfy (1), then A, B, C are three interior angles of a triangle or form the trihedral angles, dihedral angles of a parallelepiped.

An equivalent form of the Embedding Triangle Inequality is as follows.

Theorem 2. Suppose that $A + B + C = (2k + 1)\pi$, x, y, $z \in \mathbf{R}$, then

(1) $xy\sin^2\dfrac{C}{2} + zx\sin^2\dfrac{B}{2} + yz\sin^2\dfrac{A}{2} \geqslant \dfrac{1}{4}(2xy + 2yz + 2xz - x^2 - y^2 - z^2)$, equality holds if and only if $x : y : z = \sin A : \sin B : \sin C$.

(2) $(x + y + z)^2 \geqslant 4\left(xy\cos^2\dfrac{C}{2} + zx\cos^2\dfrac{B}{2} + yz\cos^2\dfrac{A}{2}\right)$,

equality holds if and only if $x : y : z = \sin A : \sin B : \sin C$.

(3) $(x + y + z)^2 \geqslant 4(yz\sin^2 A + zx\sin^2 B + xy\sin^2 C)$, equality holds if and only if $x : y : z = \sin 2A : \sin 2B : \sin 2C$.

Proof. (1) It can be proved by applying double angle formula $\cos A = 1 - 2\sin^2(A/2)$ to (1).

(2) It can be proved by applying double angle formula $\cos A = 2\cos^2(A/2) - 1$ to (1).

(3) It can be proved by applying half angle formula $\sin^2 A = (1 - \cos 2A)/2$ to (1). □

Remark. Applying sine law to (3), we can get: in $\triangle ABC$, λ, μ, $\upsilon \in \mathbf{R}$, we have

$$(\lambda + \mu + \upsilon)^2 R^2 \geqslant \mu\upsilon a^2 + \nu\lambda b^2 + \lambda\mu c^2$$
$$\Leftrightarrow (\lambda + \mu + \upsilon)^2 (abc)^2 \geqslant 16\Delta^2 (\mu\upsilon a^2 + \nu\lambda b^2 + \lambda\mu c^2) \tag{2}$$

in (2) let $\lambda = xa^2$, $\mu = yb^2$, $\upsilon = zc^2$, x, y, $z \in \mathbf{R}$, we have

$$(xa^2 + yb^2 + zc^2)^2 \geqslant 16\Delta^2 (xy + yz + zx), \tag{3}$$

equality holds if and only if

$$\lambda : \mu : \upsilon = (b^2 + c^2 - a^2) : (c^2 + a^2 - b^2) : (a^2 + b^2 - c^2).$$

Inequality (3) has wild applications, it often appears in the study of the elementary geometric inequality (for example Mr. Shan Zun's "Geometric Inequality". Shanghai Education Press, 1980).

There are other researchers who also obtained some algebra inequality more general than (3) as follows.

Example 1. Assume that there are at least two positive number among λ_1, λ_2 and λ_3, and satisfy $\lambda_1\lambda_2 + \lambda_2\lambda_3 + \lambda_3\lambda_1 > 0$, x, y, z are arbitrarily real numbers. Show that,

$$(\lambda_1 x + \lambda_2 y + \lambda_3 z)^2$$
$$\geqslant (\lambda_1\lambda_2 + \lambda_2\lambda_3 + \lambda_3\lambda_1)(2xy + 2yz + 2zx - x^2 - y^2 - z^2)$$

equality holds if and only if $x/(\lambda_2 + \lambda_3) = y/(\lambda_1 + \lambda_3) = z/(\lambda_1 + \lambda_2)$.

Answer 1. Applying embedding inequality. Since there are at least two positive numbers among λ_1, λ_2, λ_3, and $\lambda_1(\lambda_2+\lambda_3)+\lambda_2\lambda_3>0$, we obtain $\lambda_2+\lambda_3>0$. Similarly we can get $\lambda_1+\lambda_2>0$, $\lambda_1+\lambda_3>0$.

Suppose that $\lambda_1+\lambda_2=c^2$, $\lambda_2+\lambda_3=a^2$, $\lambda_1+\lambda_3=b^2(a, b, c>0)$, then

$$\lambda_1=\frac{1}{2}(b^2+c^2-a^2),\ \lambda_2=\frac{1}{2}(a^2+c^2-b^2),\ \lambda_3=\frac{1}{2}(a^2+b^2-c^2).$$

By $\lambda_1\lambda_2+\lambda_2\lambda_3+\lambda_3\lambda_1>0$, we can get

$$(a+b+c)(a-b+c)(a+b-c)(-a+b+c)>0.$$

Thus, a, b, c can form three sides of a triangle, say $\triangle ABC$. We can get that the primitive inequality

$$\Leftrightarrow\sum\lambda_1^2x^2+2\sum\lambda_1\lambda_2xy\geqslant(\sum\lambda_1\lambda_2)(2xy+2yz+2zx-x^2-y^2-z^2)$$

$$\Leftrightarrow\sum x^2(\lambda_1+\lambda_2)(\lambda_1+\lambda_3)\geqslant\sum\lambda_1\lambda_2(2yz+2xz)$$

$$\Leftrightarrow\sum x^2c^2b^2\geqslant\sum\frac{c^4-a^4-b^4+2a^2b^2}{4}(2yz+2xz)$$

$$\Leftrightarrow\sum(xcb)^2\geqslant\sum yza^2(b^2+c^2-a^2). \qquad \text{(a)}$$

In fact, by embedding inequality, we can get that

$$\begin{aligned}\sum(xbc)^2&\geqslant\sum 2(yca)(zba)\cdot\cos A\\&=\sum 2(yca)(zba)\cdot\frac{b^2+c^2-a^2}{2bc}\\&=\sum yza^2(b^2+c^2-a^2).\end{aligned}$$

This is formula (a). Therefore the primitive inequality holds.

Equality holds if and only if

$$\begin{aligned}&\frac{xbc}{\sin A}=\frac{yac}{\sin B}=\frac{zab}{\sin C}\\&\Leftrightarrow\frac{x}{\sin^2 A}=\frac{y}{\sin^2 B}=\frac{z}{\sin^2 C}\\&\Leftrightarrow\frac{x}{\lambda_2+\lambda_3}=\frac{y}{\lambda_1+\lambda_3}=\frac{z}{\lambda_1+\lambda_2}.\end{aligned}$$

□

Answer 2. Discriminant Method. By the same method to the previous, we obtain $\lambda_2+\lambda_3>0$, $\lambda_1+\lambda_2>0$, $\lambda_1+\lambda_3>0$. Now, the primitive inequality equivalent to the statement that

$$(\lambda_1+\lambda_2)(\lambda_1+\lambda_3)x^2-2[\lambda_3(\lambda_1+\lambda_2)y+\lambda_2(\lambda_1+\lambda_3)z]x+ [(\lambda_1+\lambda_2)(\lambda_2+\lambda_3)y^2+(\lambda_1+\lambda_3)(\lambda_2+\lambda_3)z^2-2\lambda_1(\lambda_2+\lambda_3)yz]\geqslant 0.$$

The expression on the left side of the previous formula is a quadratic function with respect to x, whose coefficient of the quadratic term is positive.

$$\Delta\leqslant 0$$

that is

$$\begin{aligned}&\lambda_3^2(\lambda_1+\lambda_2)^2y^2+\lambda_2^2(\lambda_1+\lambda_3)^2z^2+2\lambda_3\lambda_2(\lambda_1+\lambda_2)(\lambda_1+\lambda_3)yz\\ \leqslant\ &(\lambda_1+\lambda_2)(\lambda_1+\lambda_3)[(\lambda_1+\lambda_2)(\lambda_3+\lambda_2)y^2+(\lambda_1+\lambda_3)(\lambda_3+\lambda_2)z^2-\\ &2\lambda_1(\lambda_3+\lambda_2)yz]\\ \Leftrightarrow\ &(\lambda_1\lambda_2+\lambda_2\lambda_3+\lambda_3\lambda_1)[(\lambda_1+\lambda_2)y-(\lambda_1+\lambda_3)z]^2\geqslant 0,\end{aligned}$$

which holds by the conditions of the problem. Thus, the primitive inequality follows immediately. And equality holds if and only if

$$(\lambda_1+\lambda_2)y=(\lambda_1\lambda_3)z\Leftrightarrow\frac{y}{\lambda_1+\lambda_3}=\frac{z}{\lambda_1+\lambda_2}.$$

By the symmetry of x, y, z, we can see that equality holds if and only if

$$\frac{x}{\lambda_2+\lambda_3}=\frac{y}{\lambda_1+\lambda_3}=\frac{z}{\lambda_1+\lambda_2}.$$

□

The Answer 1 of Example 1 was provided by student Xiang Zhen. The Answer 2 was by Mr. Huang Zhiyi (the former student of the High School Affiliated to South China Normal University, who won a gold medal in the 45th IMO).

Next, we will introduce applications of the embedding inequality.

Example 2. Let the three sides of $\triangle ABC$ and $\triangle A'B'C'$ be a, b, c and a', b', c', respectively. And t_a, t_b, t_c and t'_a, t'_b, t'_c are the internal bisectors, respectively. Show that

$$t_a t'_a + t_b t'_b + t_c t'_c \leqslant \frac{3}{4}(aa' + bb' + cc'). \tag{4}$$

Proof. By internal bisector formula, we can get that

$$t_a = \frac{2bc}{b+c} \cdot \cos\frac{A}{2} \leqslant \sqrt{bc}\cos\frac{A}{2}.$$

Similarly $t'_a \leqslant \sqrt{b'c'}\cos\frac{A}{2}$, and so on. Therefore

$$\begin{aligned} t_a t'_a + t_b t'_b + t_c t'_c &\leqslant \sum \sqrt{bb'cc'} \cdot \cos\frac{A}{2}\cos\frac{A'}{2} \\ &= \frac{1}{2}\sum \sqrt{bb'cc'}\left(\cos\frac{A-A'}{2} + \cos\frac{A+A'}{2}\right) \\ &\leqslant \frac{1}{2}\sum \sqrt{bb'cc'}\left(1 + \cos\frac{A+A'}{2}\right) \\ &= \frac{1}{2}\sum \sqrt{bb'cc'} + \frac{1}{2}\sum \sqrt{bb'cc'} \cdot \cos\frac{A+A'}{2}. \end{aligned} \tag{a}$$

Now, we notice that $(A + A') + (B + B') + (C + C') = 2\pi$, applying embedding inequality and let $x = \sqrt{aa'}$, $y = \sqrt{bb'}$, $z = \sqrt{cc'}$. We obtain

$$2\sum \sqrt{bb'cc'}\cos\frac{A+A'}{2} \leqslant \sum aa'. \tag{b}$$

By mean value inequality, we obtain

$$\sum \sqrt{bb'cc'} \leqslant \frac{1}{2}\sum(bb' + cc') = \sum aa'. \tag{c}$$

By Inequalities (a), (b) and (c), we obtain Inequality (4). □

Example 3. Suppose that point P on $\triangle ABC$. Denote $PA = x$, $PB = y$ and $PC = z$. Show that

$$x^2+y^2+z^2 \geqslant \frac{1}{3}(a^2+b^2+c^2).$$

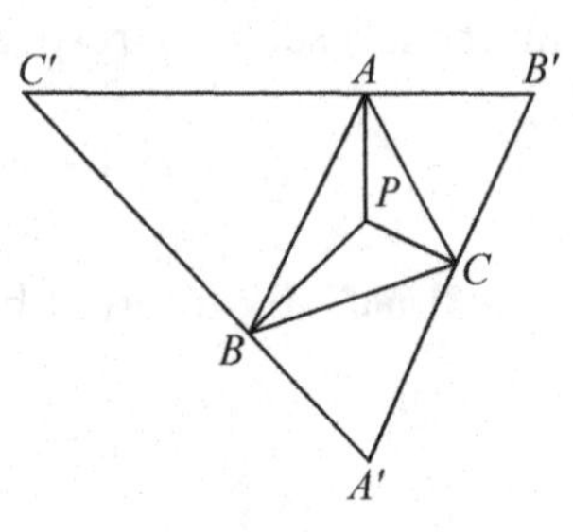

Figure 8.1

Proof. First we consider the case that P is in $\triangle ABC$. Draw three lines from the vertexes A, B and C of $\triangle ABC$ perpendicular to the segment PA, PB and PC, respectively, see Figure 8.1. The three lines are crossing pairwise at A', B' and C'. So, $\angle BPC = \pi - A'$, $\angle APB = \pi - C'$, $\angle APC = \pi - B'$. By cosine law, we can get that

$$\begin{aligned}
a^2 &= y^2+z^2+2yz\cos A', \\
b^2 &= x^2+z^2+2xz\cos B', \\
c^2 &= x^2+y^2+2xy\cos C'.
\end{aligned}$$

Adding up and using the embedding inequality, we obtain

$$\begin{aligned}
a^2+b^2+c^2 &= 2(x^2+y^2+z^2)+2xy\cos C'+2xz\cos B'+2yz\cos A' \\
&\leqslant 2(x^2+y^2+z^2)+(x^2+y^2+z^2) \\
&= 3(x^2+y^2+z^2),
\end{aligned}$$

as required. For the case that P is on the side of $\triangle ABC$, the proof is similar and is left to the reader. □

We will prove inequality of weighted sine sum, which is a useful inequality set up by Mr. Yang Xuezhi in 1988, by the embedding inequality.

Theorem 3 (Inequality of weighted sine sum). For arbitrary real numbers x, y, z, arbitrary $\triangle ABC$ and positive numbers u, v, w, we have

$$2(yz\sin A+zx\sin B+xy\sin C) \leqslant \left(\frac{x^2}{u}+\frac{y^2}{v}+\frac{z^2}{w}\right)\sqrt{vw+wu+uv}.$$

Equality holds if and only if $x:y:z=\cos A:\cos B:\cos C$ and $u:v:w=\cot A:\cot B:\cot C$.

Proof. Let $x'=\frac{x}{\sqrt{u}}$, $y'=\frac{y}{\sqrt{v}}$, $z'=\frac{z}{\sqrt{w}}$. The primitive inequality is equivalent to

$$2\sqrt{\frac{vw}{uv+vw+uw}}\cdot y'z'\sin A+2\sqrt{\frac{uw}{uv+vw+uw}}\cdot x'z'\sin B$$
$$+2\sqrt{\frac{uv}{uv+vw+uw}}\cdot x'y'\sin C$$
$$\leqslant x'^2+y'^2+z'^2. \tag{a}$$

By Cauchy inequality, the expression on the left side of the formula (a) is less than or equal to

$$2\sqrt{\frac{vw}{uv+vw+uw}+\frac{uw}{uv+vw+uw}+\frac{uv}{uv+vw+uw}}\cdot$$
$$\sqrt{y'^2z'^2\sin^2A+x'^2z'^2\sin^2B+x'^2y'^2\sin^2C}$$
$$=2\sqrt{y'^2z'^2\sin^2A+x'^2z'^2\sin^2B+x'^2y'^2\sin^2C},$$

thus, it suffices to prove that

$$2\sqrt{y'^2z'^2\sin^2A+x'^2z'^2\sin^2B+x'^2y'^2\sin^2C}\leqslant x'^2+y'^2+z'^2$$
$$\Leftrightarrow 2y'^2z'^2(2\sin^2A-1)+2x'^2z'^2(2\sin^2B-1)+2x'^2y'^2(2\sin^2C-1)$$
$$\leqslant x'^4+y'^4+z'^4$$
$$\Leftrightarrow 2y'^2z'^2\cos(\pi-2A)+2x'^2z'^2\cos(\pi-2B)+2x'^2y'^2\cos(\pi-2C)$$
$$\leqslant x'^4+y'^4+z'^4. \tag{b}$$

Formula (b) can be obtained by the substitutions $(x, y, z)\rightarrow(x'^2, y'^2, z'^2)$ and $(A, B, C)\rightarrow(\pi-2A, \pi-2B, \pi-2C)$ by the embedding inequality. □

By applying the inequality of weighted sine sum, we can obtain the following two interesting triangle inequalities.

Example 4. Let three sides of $\triangle ABC$ be a, b, c, the corresponding angles be A, B, C. Denotes $s=a+b+c$. And let three sides of $\triangle A'$, B', C' be a', b', c', the corresponding angle be A', B', C'. Denote $s=a+b+c$. Show that

$$\frac{a}{a'}\tan\frac{A}{2}+\frac{b}{b'}\tan\frac{B}{2}+\frac{c}{c'}\tan\frac{C}{2}\geqslant\frac{\sqrt{3}s}{2s'}.$$

Equality holds if and only if $\triangle ABC$ and $\triangle A'B'C'$ are regular triangles.

Proof. Substituting in the inequality of weighted sine sum by $A \to \frac{\pi - A}{2}$, $B \to \frac{\pi - B}{2}$, $C \to \frac{\pi - C}{2}$, we obtain a new inequality, then substituting in the new inequality by $x \to yz$, $y \to zx$, $z \to xy$. And substituting $u \to a'$, $v \to b'$, $w \to c'$, obtain

$$\frac{yz}{a'x} + \frac{xz}{b'y} + \frac{xy}{c'z} \geqslant \frac{2\left(x\cos\frac{A}{2} + y\cos\frac{B}{2} + z\cos\frac{C}{2}\right)}{\sqrt{b'c' + a'c' + a'b'}}. \tag{a}$$

And in (a), we use the substitutions $x \to \frac{bc}{s}\cos\frac{A}{2}$, $y \to \frac{ac}{s}\cos\frac{B}{2}$, $z \to \frac{ab}{s}\cos\frac{C}{2}$, and notice that

$$\frac{\frac{a}{s}\cos\frac{B}{2}\cos\frac{C}{2}}{\cos\frac{A}{2}} = \frac{2R\sin A\cos\frac{B}{2}\cos\frac{C}{2}}{2R(\sin A + \sin B + \sin C)\cos\frac{A}{2}}$$

$$= \frac{2\sin\frac{A}{2}\cos\frac{A}{2}\cos\frac{B}{2}\cos\frac{C}{2}}{4\cos\frac{A}{2}\cos\frac{B}{2}\cos\frac{C}{2}\cos\frac{A}{2}} = \frac{1}{2}\tan\frac{A}{2},$$

and so on. So we can obtain

$$\frac{1}{2}\left(\frac{a}{a'}\tan\frac{A}{2} + \frac{b}{b'}\tan\frac{B}{2} + \frac{c}{c'}\tan\frac{C}{2}\right) \geqslant \frac{2\left(\frac{bc}{s}\cos^2\frac{A}{2} + \frac{ac}{s}\cos^2\frac{B}{2} + \frac{ab}{s}\cos^2\frac{C}{2}\right)}{\sqrt{b'c' + a'c' + a'b'}}. \tag{b}$$

Since

$$b'c' + a'c' + a'b' \leqslant \frac{4}{3}s'^2 \tag{c}$$

and

$$
\begin{aligned}
&2\left(\frac{bc}{s}\cos^2\frac{A}{2}+\frac{ac}{s}\cos^2\frac{B}{2}+\frac{ab}{s}\cos^2\frac{C}{2}\right)\\
&=\frac{1}{s}[bc(1+\cos A)+ac(1+\cos B)+ab(1+\cos C)]\\
&=\frac{1}{2s}[2bc+(a^2+b^2-c^2)+2ac+(a^2+c^2-b^2)+2ab+(-a^2+b^2+c^2)]\\
&=\frac{1}{2s}[a^2+b^2+c^2+2(ab+bc+ca)]\\
&=\frac{1}{2s}(a+b+c)^2=\frac{s}{2}. \qquad \text{(d)}
\end{aligned}
$$

Substitute (c) and (d) into (b), the inequality follows immediately.

□

By embedding inequality, we can get well-known Neuberg-Pedoe's inequality and some inequalities about moving point inside the triangle. Limited to the aim of this book, we do not expand the subject in this area any further.

Another well-known inequality with weighted real numbers is the inequality for moment of inertia.

Theorem 4 (Inequality for moment of inertia). Let P be an arbitrary point on the plane of $\triangle ABC$. Denote $PA=R_1$, $PB=R_2$, $PC=R_3$, then for arbitrary real numbers $x, y, z \in \mathbf{R}$, we have

$$(x+y+z)(xR_1^2+yR_2^2+zR_3^2) \geqslant yza^2+xzb^2+xyc^2. \qquad (5)$$

Equality holds if and only if $xR_1 : yR_2 : zR_3 = \sin\alpha_1 : \sin\alpha_2 : \sin\alpha_3$, which $\alpha_i = \angle A_{i+1}PA_{i+2}$ ($A_4=A_1$, $A_5=A_2$, $i=1, 2, 3$) (the angles are in the same direction).

Proof. Let $\overrightarrow{PQ}=\dfrac{\sum x\overrightarrow{PA}}{\sum x}$, then

$$
\begin{aligned}
0 &\leqslant (\textstyle\sum x)^2 \mid \overrightarrow{PQ} \mid^2 = \mid \textstyle\sum x\overrightarrow{PA} \mid^2\\
&= \textstyle\sum x^2 \mid \overrightarrow{PA} \mid^2 + 2\textstyle\sum xy\,\overrightarrow{PA}\cdot\overrightarrow{PB}
\end{aligned}
$$

$$= \sum x^2 \mid \overrightarrow{PA} \mid^2 + \sum xy(\mid \overrightarrow{PA} \mid^2 + \mid \overrightarrow{PB} \mid^2 - \mid \overrightarrow{AB} \mid^2)$$
$$= (\sum x)(\sum x \mid \overrightarrow{PA} \mid^2) - \sum xy \mid \overrightarrow{AB} \mid^2$$
$$= (\sum x)(\sum x R_1^2) - \sum xyc^2.$$

Thus, Inequality (5) follows immediately. □

Remark. If point P is the interior point of $\triangle ABC$, the conditions for that the equality holds are

$$\frac{ar_1}{x} = \frac{br_2}{y} = \frac{cr_3}{z},$$

where r_1, r_2, r_3 is the distance form point P to lines BC, AC, AB, respectively.

Example 5 (Klamkin's inequality). Let two arbitrary points P and P' on the plane of $\triangle ABC$, and denote $PA_i = R_i$, $P'A_i = R'_i$, three sides be $a_i = A_{i-1}A_{i+1}$, where $A_4 = A_1$, $A_0 = A_3$, $i = 1, 2, 3$. Show that

$$a_1 R_1 R'_1 + a_2 R_2 R'_2 + a_3 R_3 R'_3 \geqslant a_1 a_2 a_3, \tag{6}$$

and point out the conditions such that the equality holds.

Proof. In Theorem 4, let $x = a_1 R'_1 / R_1$, $y = a_2 R'_2 / R_2$ and $z = a_3 R'_3 / R_3$, we can get

$$\left(\sum \frac{a_i R'_i}{R_i}\right)\left(\sum a_i R'_i R_i\right) \geqslant \sum a_1^2 \left(\frac{a_2 R'_2}{R_2}\right)\left(\frac{a_3 R'_3}{R_3}\right).$$

By simplifying above inequality yields

$$(\sum a_1 R'_1 R_2 R_3)(\sum a_1 R_1 R'_1) \geqslant a_1 a_2 a_3 (\sum a_1 R_1 R'_2 R'_3). \tag{a}$$

Similarly,

$$(\sum a_1 R_1 R'_2 R'_3)(\sum a_1 R_1 R'_1) \geqslant a_1 a_2 a_3 (\sum a_1 R'_1 R_2 R_3). \tag{b}$$

Add two equations then divide both sides by the same terms. The inequality is derived.

Notice that in formula (a) the equality holds if

$$\frac{r_1R_1}{R_1'}=\frac{r_2R_2}{R_2'}=\frac{r_3R_3}{R_3'}. \tag{c}$$

In formula (b) the equality holds if

$$\frac{r_1'R_1'}{R_1}=\frac{r_2'R_2'}{R_2}=\frac{r_3'R_3'}{R_3} \tag{d}$$

where r_i, r'_i is the distance form point P and P' to a_i, respectively.

Multiply formula (c) and (d), we obtain

$$r_1r_1'=r_2r_2'=r_3r_3'.$$

The relation is inverse ratio between the distance from point P to three sides and the distance from point P' to three sides. This shows that P and P' are a pair of isogonal conjugate points to $\triangle A_1A_2A_3$. □

Remark. (1) The concept and nature of isogonal conjugate points can be found in Chapter VIII in reference book "Modern Geometry" of Roger A. Johnson.

(2) The subtle proof above is that applying the inequality of moment of inertia symmetrically.

Next inequality is established by Mr. Tang Lihua.

Example 6. Let P be an arbitrary point on the plane of $\triangle A_1A_2A_3$, with area $\triangle$ and sides a_1, a_2, a_3, then

$$\begin{aligned}&(R_2^2+R_3^2-R_1^2)\sin A_1+(R_3^2+R_1^2-R_2^2)\sin A_2\\&\quad+(R_1^2+R_2^2-R_3^2)\sin A_3\geqslant 2\Delta,\end{aligned} \tag{7}$$

where $R_i=PA_i$, $i=1, 2, 3$.

Equality holds if and only if $a_1=a_2=a_3$ and

$$R_1 : R_2 : R_3=\sin\alpha_1 : \sin\alpha_2 : \sin\alpha_3,$$

where $\angle\alpha_i=\angle A_{i+1}PA_{i+2}$ ($A_4=A_1$, $A_5=A_2$, $i=1, 2, 3$) (the angles are taken in the same direction).

We first prove the following lemma.

Lemma 1. Let $\triangle A_1A_2A_3$ be the triangle with sides a_1, a_2, and a_3, then

$$a_1(a_1+a_2-a_3)(a_1+a_3-a_2)+a_2(a_2+a_1-a_3)(a_2+a_3-a_1) \\ +a_3(a_3+a_1-a_2)(a_3+a_2-a_1) \geqslant 3a_1a_2a_3. \quad \text{(a)}$$

Equality holds if and only if $a_1 = a_2 = a_3$.

Proof. Let

$$x = \frac{1}{2}(a_2+a_3-a_1),\ y = \frac{1}{2}(a_3+a_1-a_2),\ z = \frac{1}{2}(a_1+a_2-a_3),$$

then $x, y, z > 0$ and

$$a_1 = y+z, \quad a_2 = z+x, \quad a_3 = x+y.$$

So Formula (a) is equivalent to

$$4[yz(y+z)+zx(z+x)+xy(x+y)] \geqslant 3(x+y)(y+z)(z+x) \\ \Leftrightarrow 4[x^2(y+z)+y^2(z+x)+z^2(x+y)] \geqslant 3[x^2(y+z) \\ +y^2(z+x)+z^2(x+y)]+6xyz \\ \Leftrightarrow x^2(y+z)+y^2(z+x)+z^2(x+y) \geqslant 6xyz. \quad \text{(b)}$$

By mean value inequality, Formula (b) is established. Thus, Formula (a) in the Lemma follows immediately. Equality holds if and only if $x = y = z$, that is $a_1 = a_2 = a_3$. □

We now prove Formula (7) as follows.

Proof. By the sine law and $\Delta = a_1a_2a_3/(4R)$, Formula (7) is equivalent to

$$(a_2+a_3-a_1)R_1^2+(a_1+a_3-a_2)R_2^2+(a_1+a_2-a_3)R_3^2 \geqslant a_1a_2a_3. \quad \text{(c)}$$

Assume that $a_1 \geqslant a_2 \geqslant a_3 > 0$, then

$$a_1+a_2-a_3 \geqslant a_3+a_1-a_2 \geqslant a_2+a_3-a_1 > 0.$$

Denote $\lambda_i = (a_1 + a_2 + a_3 - 2a_i)$ $(i = 1, 2, 3)$, so,

$$\lambda_2\lambda_3 a_1 \geqslant \lambda_3\lambda_1 a_2 \geqslant \lambda_1\lambda_2 a_3.$$

By Lemma 1 and inequality for moment of inertia and Tchebychev's inequality, we obtain

$$(a_2 + a_3 - a_1)R_1^2 + (a_1 + a_3 - a_2)R_2^2 + (a_1 + a_2 - a_3)R_3^2$$

$$= \lambda_1 R_1^2 + \lambda_2 R_2^2 + \lambda_3 R_3^2 \geqslant \frac{\sum_{i=1}^{3} \lambda_{i+1}\lambda_{i+2} \cdot a_i^2}{\sum_{i=1}^{3} \lambda_i}$$

$$= \frac{\sum_{i=1}^{3} (\lambda_{i+1}\lambda_{i+2} \cdot a_i) \cdot a_i}{\sum_{i=1}^{3} a_i} \geqslant \frac{(\sum_{i=1}^{3} \lambda_{i+1}\lambda_{i+2} \cdot a_i)(\sum_{i=1}^{3} a_i)}{3\sum_{i=1}^{3} a_i}$$

$$\geqslant a_1 a_2 a_3.$$

where $\lambda_4 = \lambda_1$, $\lambda_5 = \lambda_2$.

Thus, Formula (c) is established and Formula (7) follows. By Lemma 1 and Tchebychev's inequality, equality holds if and only if $a_1 = a_2 = a_3$ and $R_1 : R_2 : R_3 = \sin \alpha_1 : \sin \alpha_2 : \sin \alpha_3$. □

Exercises 8

1. Let a, b, c be the three sides of an arbitrary triangle, and x, y, z be arbitrary real numbers. Show that

$$a^2(x - y)(x - z) + b^2(y - x)(y - z) + c^2(z - x)(z - y) \geqslant 0.$$

2. In $\triangle ABC$ and $\triangle A'B'C'$, show that

$$\cot A + \cot B + \cot C \geqslant \frac{\cos A'}{\sin A} + \frac{\cos B'}{\sin B} + \frac{\cos C'}{\sin C}.$$

3. (Garfunkel-Baukoff)

$$\tan^2\frac{A}{2} + \tan^2\frac{B}{2} + \tan^2\frac{C}{2} \geqslant 2 - 8\sin\frac{A}{2}\sin\frac{B}{2}\sin\frac{C}{2}.$$

4. (Yang Xuezhi) Let x, y, z, $w \in \mathbf{R}^+$, $\alpha + \beta + \gamma + \theta = (2k + 1)\pi$, $k \in \mathbf{Z}$, then

$$|x\sin\alpha + y\sin\beta + z\sin\gamma + w\sin\theta| \leqslant \sqrt{\frac{(xy + zw)(xz + yw)(xw + yz)}{xyzw}}.$$

5. (Tang, Lihua) Let P be an arbitrary point on the plane of $\triangle A_1A_2A_3$ with area Δ and sides a_1, a_2, a_3, then

$$(a_2 + a_3 - a_1)R_1^2 + (a_3 + a_1 - a_2)R_2^2 + (a_1 + a_2 - a_3)R_3^2 \geqslant \frac{16}{3}\Delta^2,$$

where $R_i = PA_i$, $i = 1, 2, 3$.

Chapter 9 Locus problem of Tsintsifas's inequality

The following is a well known conclusion in Euclidean geometry.

If A, B, C, D are any four points on plane, then $AD \cdot BC$, $BD \cdot CA$ and $CD \cdot AB$ are three edges of a triangle; when A, B, C and D are on a circle, this triangle is degradation (in this case, we have the well-known Ptolemy's theorem).

The conclusion is not difficult to prove but is much easier using complex number.

In fact, let the complex numbers z_1, z_2, z_3 and z_4, correspond to A, B, C and D, respectively. Then the conclusion follows by the following identity

$$(z_1 - z_4)(z_2 - z_3) + (z_2 - z_4)(z_3 - z_1) + (z_3 - z_4)(z_1 - z_2) = 0.$$

Tsintsifas proposed the following problem on Crux. Math. (9):

Problem. Let T be a given $\triangle ABC$, P be a point in the same plane of T, denote T_0 a triangle with sides $a \cdot PA$, $b \cdot PB$, $c \cdot PC$ (may be degradation). Let R, R_0 be the circumcircle radius of T and T_0, respectively. Find the locus of point P, such that

$$PA \cdot PB \cdot PC \leqslant RR_0.$$

This is a problem full of challenge. The following answer is given by student Zhu Qinsan. The crux of the solution is to apply the pedal triangle's area formula to point P.

Answer 1. We first prove a lemma.

Lemma 1. Let P be a point on plane of $\triangle A_1A_2A_3$, then the oriented area of P to $\triangle A_1A_2A_3$'s pedal triangle is

$$F = \frac{R^2 - \overline{OP}^2}{4R^2} \cdot S_{\triangle A_1A_2A_3},$$

where O is the circumcenter of $\triangle A_1A_2A_3$.

Proof. Let $\triangle P_1P_2P_3$ be a pedal triangle and let A_2P cross the circumcircle of $\triangle A_1A_2A_3$ at point B_2, see Figure 9.1. Then

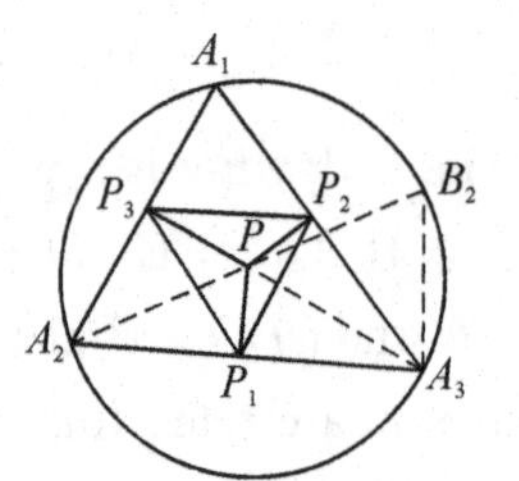

Figure 9.1

$$\angle A_2PA_3 = \angle P_2P_1P_3 + \angle A_2A_1A_3$$
$$= \angle A_2B_2A_3 + \angle B_2A_3P.$$

Note that the angle is oriented, then

$$\angle P_2P_1P_3 = \angle B_2A_3P.$$

Thus,

$$F = \frac{1}{2}\overline{P_1P_2} \cdot \overline{P_1P_3} \sin \angle P_2P_1P_3$$
$$= \frac{1}{2}\overline{P_1P_2} \cdot \overline{P_1P_3} \sin \angle B_2A_3P$$
$$= \frac{1}{2}\overline{PA_3} \cdot \sin \alpha_3 \overline{PA_2} \sin \alpha_2 \sin \angle B_2A_3P$$

for

$$\frac{\sin \angle B_2A_3P}{\sin \angle A_2B_2A_3} = \frac{\overline{PB_2}}{\overline{PA_3}}$$

then

$$F = \frac{1}{2}\overline{PA_2PB_2} \sin \angle A_2B_2A_3 \sin \alpha_2 \sin \alpha_3$$
$$= \frac{1}{2}(R^2 - \overline{OP}^2) \sin \alpha_1 \sin \alpha_2 \sin \alpha_3$$
$$= \frac{R^2 - \overline{OP}^2}{4R^2} \cdot S_{\triangle A_1A_2A_3}.$$

We now go to the original problem.

Proof. Let $PA = a'$, $PB = b'$, $PC = c'$, then,

$$R = \frac{abc}{4S_T}, \quad R_0 = \frac{abca'b'c'}{4S_{T_0}},$$

then, $RR_0 \geqslant PA \cdot PB \cdot PC$, which is equivalent to

$$\frac{(abc)^2}{16S_T S_{T_0}} \geqslant 1.$$

That is

$$S_{T_0} \leqslant \frac{(abc)^2}{16S_T}. \tag{a}$$

(See Figure 9.2.) Let A', B', C' be projections of point P on BC, CA, AB, respectively. Then

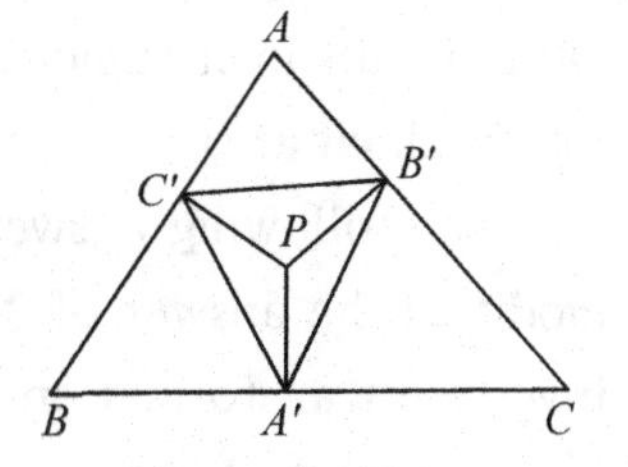

Figure 9.2

$$B'C' = PA \sin A = \frac{a \cdot PA}{2R}.$$

Similarly,

$$C'A' = \frac{b \cdot PB}{2R}, \quad A'B' = \frac{c \cdot PC}{2R},$$

then $\triangle A'B'C'$ is similar with T_0 and the similarity ratio is $2R$. Then by (a), we have

$$S_{\triangle A'B'C'} = \frac{1}{4R^2} S_{T_0} \leqslant \frac{1}{4R^2} \cdot \frac{a^2b^2c^2}{16S_T}. \tag{b}$$

Applying Lemma 1 we have

$$S_{\triangle A'B'C'} = \frac{|R^2 - \overline{OP}^2|}{4R^2} \cdot S_T. \tag{c}$$

Then by (b) and (c), we have

$$\frac{|R^2 - \overline{OP}^2|}{4R^2} \cdot S_T \leqslant \frac{1}{4R^2} \cdot \frac{a^2b^2c^2}{16S_T}.$$

This is equivalent to

$$|R^2 - \overline{OP}^2| \leqslant \frac{a^2b^2c^2}{16S_T} = R^2,$$

then,

$$|\overline{OP}| \leqslant \sqrt{2}R.$$

This explains the locus of the point P is a disk with center O (T's circumcenter) and radius $\sqrt{2}R$. □

Remark. Lemma 1 is called Gergonne's formula, which is a classical conclusion in plane geometry (see the book "Modern Geometry" by Roger A. Johnson). By Lemma 1, an interesting conclusion can be deduced: the locus of a point to which the pedal triangle with given area is a circle with the center of the circumcircle. Of all points in circumcircle, the pedal triangle to the circumcenter has the maximal area.

The following answer was given by Li Xianying (student), who modified the answer of Xiang Zhen (student). The crux is using the inversion transformation.

Answer 2. Let O be the circumcenter of $\triangle ABC$. Take P as the inversion center, and $\lambda = PA \cdot PB \cdot PC$ the inversion power.

Suppose that A', B', C' are inversion points of A, B, C, respectively. Then

$$B'C' = BC \cdot \frac{\lambda}{PB \cdot PC} = a \cdot PA$$

similarly

$$C'A' = b \cdot PB, \quad A'B' = c \cdot PC$$

then $\triangle A'B'C'$ is triangle which has the edges $a \cdot PA$, $b \cdot PB$, $c \cdot PC$.

Then circle $\odot(O, R)$ has the radius by inversion

$$R_0 = R \frac{\lambda}{|PO|^2 - R^2},$$

then $PA \cdot PB \cdot PC \leqslant RR_0$ which is equivalent to:

$$\lambda \leqslant R^2 \cdot \frac{\lambda}{|PO|^2 - R^2},$$

this is also equivalent to

$$| PO | \leqslant \sqrt{2}R.$$

This explains the locus of point P is the disk with center O and radius $\sqrt{2}R$.

Remark. (1) The answer of Xiang Zhen (student) takes O as the invention center, so the proof is tedious, while Li Xianying (student) takes P as invention center P and shorten the proof greatly.

(2) We can see from the above proof that there exits an invention image of $\triangle ABC$ similar to the pedal triangle to P.

Exercises 9

1. Let P be a moving point in the same plane of $\triangle ABC$, the area of the pedal triangle to point P is F_p, the area of $\triangle ABC$ is F, find the locus of point P such that $F_p \leqslant F/4$.

2. (Tsintsifas) Let P be an inner point of $\triangle ABC$ with sides a, b, c, $\triangle A'B'C'$ with sides a', b', c' be the pedal triangle to P. Show that

$$a'/a + b'/b + c'/c < 2.$$

3. (Tsintsifas and Klamkin) Let O and I be circumcenter and innercenter, respectively. The projective of point P to BC, AC, AB are D, E, F, respectively. r, r' are the radius of inscribed circle $\triangle ABC$ and $\triangle DEF$, respectively. Suppose that $OP \geqslant OI$, show that $r' \leqslant r/2$.

Chapter 10 Shum's minimal circle problem

In this chapter, I'd like to give an introduction to the problem of extremum counting in the combinatorial geometry. This was an open problem at that time put forward by George F. Shum on the Journal of Amer. Math. Monthly (1978, 824, E2746).

The problem is as follows: Suppose that $\tau = \{A_1, A_2, \ldots, A_n\}$ is the set of n non-collinear points on plane. If a circle with center O and radius r, passes through at least three points of τ, and $A_kO \leqslant r$ for all $k \in \{1, 2, \ldots, n\}$, then we call this circle is the minimal circle of the point set τ. The problem is: for fixed value n, how many minimal circles of τ are there at most?

Here, we give out three answers, of which Answer 1 is due to Strauss.

Answer 1. The answer to this problem is $n - 2$.

First, we should note that the point A_i appearing on the minimal circle must be on the boundary of the convex hull of $\{A_i\}$. Therefore, we can suppose that these n points are the n vertexes of a convex polygon P_n.

Then suppose that points A_i and A_j are on the minimal circle C_1, while A_k and A_h are on the minimal circle C_2. If line segments A_iA_j and A_kA_h intersect (and the two have common inner point), we have $C_1 = C_2$. To see this, it suffices to consider the case of the convex quadrilateral $Q = A_iA_kA_jA_h$. A_sQ is within C_1, we get

$$\angle A_kA_jA_h + \angle A_hA_iA_k \leqslant 180°, \tag{a}$$

Q is within C_2 as well, so we get

$$\angle A_iA_kA_j + \angle A_jA_hA_i \leqslant 180°. \qquad \text{(b)}$$

If we add up (a) and (b), we get the result that the sum of the four internal angles of Q is equal or less than 360°. Therefore, we can infer that equalities must hold for both (a) and (b). That is to say, Q is an inscribed quadrilateral of a circle.

So we conclude that the edges of the triangles determining the different minimal circles have no point of intersection (no common inner point). Since P_n can be divided into at most $n-2$ triangles which have no common inner point and whose vertexes are the original ones of P_n, we infer that P_n has at most $n-2$ minimal circles.

Now let us prove that $n-2$ minimal circles are attainable.

By induction, we construct an n-elements point set $\{A_k\}$, ensuring all its minimal circles are circumcircles C_k ($k = 3, \ldots, n$) which pass through $A_1A_{k-1}A_k$. Then we choose three non-collinear points $A_1A_2A_3$. Suppose that $A_1, \ldots, A_k$ ($k \geqslant 3$) have already been specified, and their minimal circles are $C_3, \ldots, C_k$. Then as Figure 10.1 indicates, we choose an inner point as A_{k+1} in the region S_k, which is formed by the chord A_1A_k and the arc of the circle C_k not containing the point A_{k-1}. Thus the circle C_{k+1} includes S_k's complement set S_k' in the circle C_k. Or actually it includes all the points $A_1, \ldots, A_{k+1}$. On the other hand, we have:

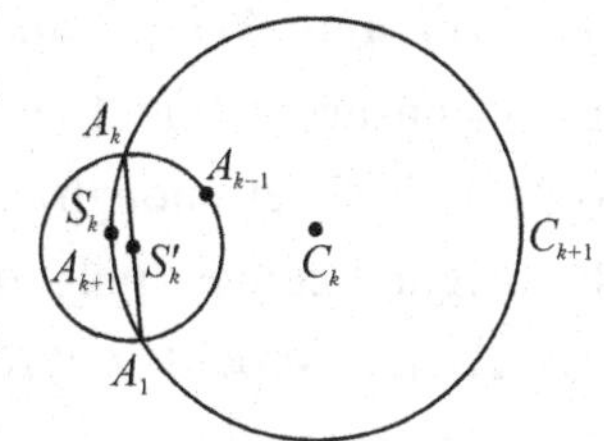

Figure 10.1

$$S_3 \supset S_4 \supset \cdots \supset S_k.$$

Therefore, we get the result that A_k are all located within the circle C_i ($i = 1, 2, \ldots, k$). In this way, our aim is attained. In fact, it is not difficult to infer that any convex polygon with n edges has $n-2$ minimal circles, as long as its four vertexes are not on the same circle.

Now let us turn to the answer given by Zhu Qinsan (student).

Answer 2. There are at most $n - 2$ minimal circles.

First we need to prove by induction that such a convex polygon with n edges that has at least $n - 2$ minimal circles does exist.

The above claim is obviously true when n is 3.

Suppose that it is also true when n is k, that is, a convex polygon with k edges $A_1A_2 \cdots A_k$ has $k - 2$ minimal circles. Now as Figure 10.2 shows, we prolong the edges A_2A_1 and $A_{k-1}A_k$, and locate a point A_{k+1} in the intersection region formed by the inboard of A_1A_2 and A_kA_{k-1} and the outboard of A_1A_k, ensuring that the distance from A_{k+1} to A_1A_k is short enough, and that A_{k+1} is within any minimal circle of the original convex polygon with k edges. Besides, the circumcircle of $\triangle A_1A_{k+1}A_k$ should contain all the A_i ($i = 2, \ldots, k - 1$. So it suffices to ensure that $\angle A_1A_{k+2}A_k$ is greater than all the supplementary angles of $\angle A_1A_iA_k$). In this case, the convex polygon with $k + 1$ edges $A_1A_2 \cdots A_kA_{k+1}$ has at least $k - 1$ minimal circles.

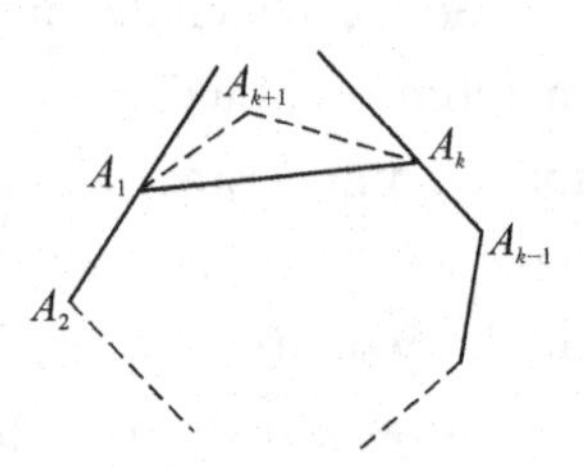

Figure 10.2

Next, we need to prove that τ has at most $n - 2$ minimal circles.

First, suppose that τ constructs a convex polygon with n edges, that is $P_n = A_1A_2 \cdots A_n$ (if it were not the case, we could think of its convex hull set τ'. Obviously, the points on the minimal circle must be on the convex hull), and suppose that no four vertexes are on the same circle.

Now, we need to give four lemmas before going to the claim.

Lemma 1. Each edge of P_n has one and only one minimal circle which passes through its two vertexes.

Proof. As Figure 10.3 shows, except A_1, A_2, all points of τ are on the same side of the edge A_1A_2. If the circumcircle of $\triangle A_1A_2A_k$ is the minimal circle, it can be inferred that A_i is within $\odot A_1A_2A_k$ and is on the same side of A_1A_2 with A_k. So $\angle A_1A_kA_2$ is the smallest of all the $\angle A_1A_iA_2$ $(i = 3, \ldots, n)$. That is to say, there can be only one minimal circle which passes through A_1A_2. In this way, Lemma 1 is proved. □

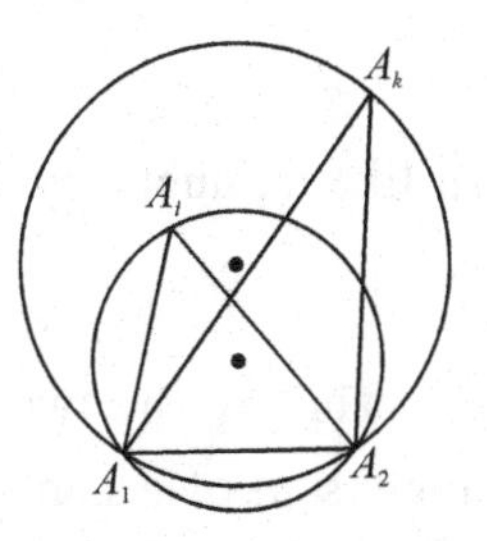

Figure 10. 3

Lemma 2. Of each diagonal of P_n, there are two or no minimal circle passing through its two vertexes.

Proof. As Figure 10. 4 shows, suppose that on one side of the diagonal A_1A_k, $\angle A_1A_iA_k$ is the smallest of all the angles whose vertexes are those of P_n, and on the other side of the diagonal, $\angle A_1A_jA_k$ is the smallest. Then if $\angle A_1A_iA_k + \angle A_1A_jA_k < \pi$, it is inferred that there is no coverage circle passing through A_1A_k. If, however, $\angle A_1A_iA_k + \angle A_1A_jA_k > \pi$, we get that the circumcircles of $\triangle A_1A_iA_k$ and $\triangle A_1A_jA_k$ are both minimal circles. Therefore, Lemma 2 is proved. □

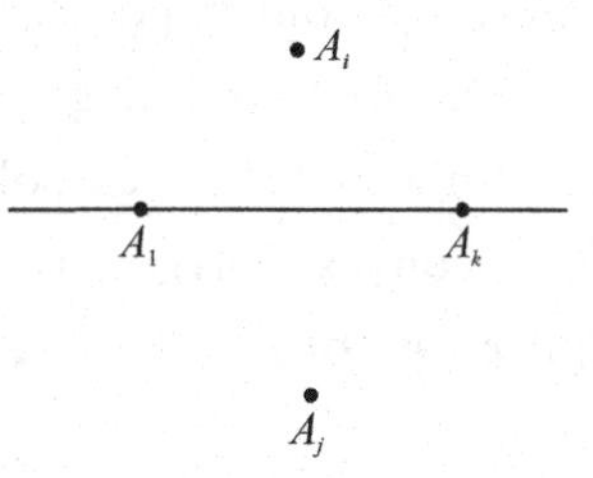

Figure 10. 4

Lemma 3. If there is a minimal circle passing through the two vertexes of the diagonal A_1A_k, A_1A_k is then called the good diagonal. And the good diagonals of P_n will not intersect at places other than the vertexes.

Proof. As Figure 10. 5 shows, suppose that the good diagonals A_iA_j and A_rA_s intersect at places that were not vertexes.

As the points A_r and A_s were both within the minimal circle which passes through A_iA_j, and the four points $A_iA_rA_s$ and A_j were not on the same circle, therefore:

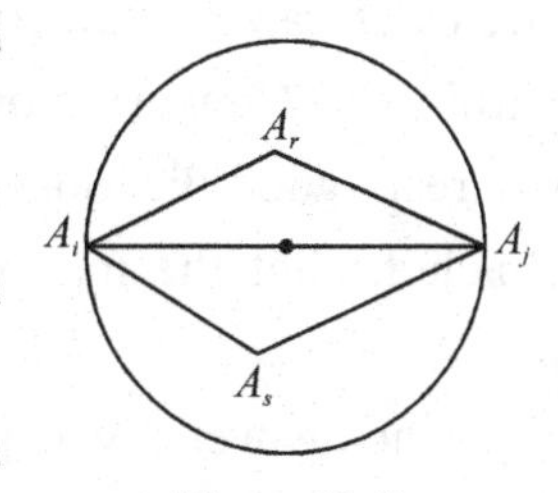

Figure 10. 5

$$\angle A_iA_rA_j + \angle A_iA_sA_j > \pi.$$

In like manner, we would get:

$$\angle A_rA_iA_s + \angle A_rA_jA_s > \pi.$$

Having the two added up together, we get the conclusion that the angle sum of a convex quadrilateral were greater than 2π, which is a contradiction. □

Lemma 4. P_n has at most $n - 3$ "good diagonals".

Proof. Suppose there are k "good diagonals", then P_n is divided into $k + 1$ convex polygons.

As the "good diagonals" do not intersect within P_n, the angle sum of all the $k + 1$ convex polygons equals that of P_n, and the angle sum of each convex polygon is equal or greater than π. As we know, the angle sum of P_n is $(n - 2)\pi$. So we get:

$$(k + 1)\pi \leqslant (n - 2)\pi,$$

that is: $k \leqslant n - 3$.

Now let us prove that the number of P_n's minimal circles is equal or lesser than $n - 2$.

As each minimal circle happens to have three chords whose ends are the vertexes of P_n, and each chord is either the edge or the good diagonal of P_n, we get:

$$\text{The number of minimal circles} \leqslant \frac{n + 2(n - 3)}{3} = n - 2.$$

If, however, there are four points of P_n that are on the same circle, it can be inferred that two of its "good diagonals" intersect within P_n. Now erase one of the "good diagonals", then the number of the rest "good diagonals" is less than $n - 2$. In this way, we can also conclude that the number of the minimal circles is equal or less than $n - 2$. □

The answer 3 was provided by student Long Yun.

Answer 3. τ has at most $n-2$ minimal circles.

Proof. Let us prove it by induction.

(i) When $n=3$, $\tau'=\{A_1, A_2, A_3\}$ happens to have one minimal circle.

(ii) Suppose this claim also exists when $n=k$. And we can have the following assumptions: For $\tau''=\{A_1, A_2, \ldots, A_{k+1}\}$, $\odot O$ is the one whose radius is the greatest of all its minimal circles. And the three points $A_1A_2A_3$ of τ'' are on $\odot O$, while the rest points of τ'' are within or on $\odot O$.

We may as well assume that $\angle A_1$ is the greatest in $\triangle A_1A_2A_3$, then $\angle A_2$ and $\angle A_3$ are both acute angles. Now we need to prove that there is no minimal circle passing through A_1 except $\odot O$. If this is not the case, and if $\odot K$ does pass through A_1, it can be inferred that A_2A_3 are on or within $\odot K$, since the minimal circle covers all the points. In fact, at least one of the two points is within $\odot K$, and we can suppose that the point is A_2. Then prolong A_2A_3, and make it intersect with $\odot K$ at B and C, as Figure 10.6 shows. So we get:

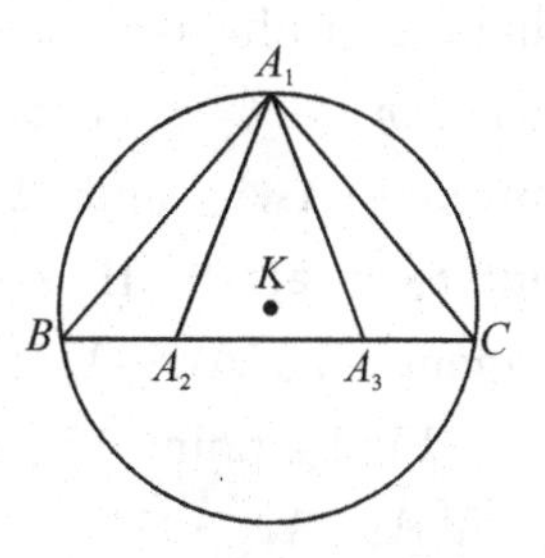

Figure 10.6

$$\angle BA_2A_1 > 90^\circ.$$

Therefore, $A_1B > A_1A_2$, and $\angle A_1CA_3 \leqslant \angle A_1A_3A_2$. That is to say:

$$\text{The radius of } \odot K = \frac{A_1B}{2\sin\angle A_1CA_2} > \frac{A_1A_2}{2\sin\angle A_1A_3A_2}$$

$$= \text{the radius of } \odot O.$$

A contradiction! Therefore, there is only one minimal circle passing through A_1.

Therefore, the minimal circles of τ'' not passing through A_1 are all the minimal circles of $\{A_2, \ldots, A_{k+1}\}$. By induction we know there are at most $k-2$ circles of this kind. Therefore, τ'' has at most $k-1$ minimal circles.

In this way, we prove that the number of minimal circles of τ is equal or less than $n-2$.

Still we need to prove that for any $n \geqslant 3$, $n \in \mathbf{N}$, the point set $\tau = \{A_1, A_2, \ldots, A_n\}$ exists, which happens to have $n-2$ minimal circles. We can view $A_1A_2 \cdots A_n$ as a convex polygon with n edges. And we will resort to the inductive method here as well.

(i) When $n=3$, the claim is attainable by choosing the three vertexes of a triangle.

(ii) When $n=k$, suppose there is such a convex polygon with k edges $A_1A_2 \cdots A_k$ as happens to have $k-2$ minimal circles. Then when $n=k+1$, as Figure 10.7 shows, prolong A_2A_1 and $A_{k-1}A_k$, making them intersect at B. If α is the smallest angle among all $\angle A_1A_iA_k (i=2, 3, \ldots, k-1)$.

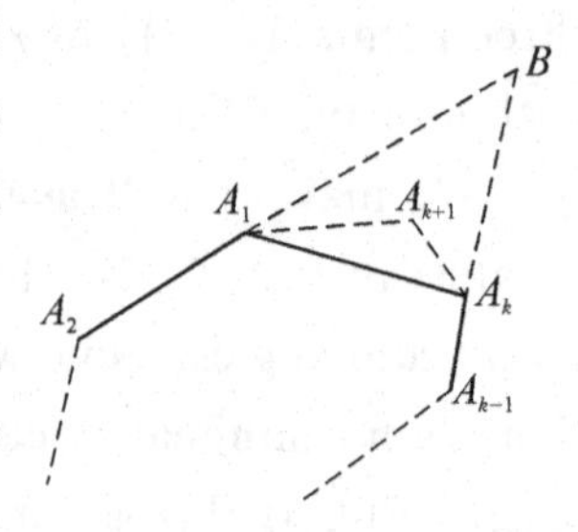

Figure 10.7

Find a point A_{k+1} in $\triangle BA_1A_k$, ensuring $\angle A_1A_{k+1}A_k > 180° - \alpha$, In this case, the $k-2$ minimal circles of $A_1A_2 \cdots A_k$ are also the minimal circles of $A_1A_2\cdots A_{k+1}$. And the circumcircle of $\triangle A_{k+1}A_1A_k$ is a new minimal circle. Therefore, $A_1A_2\cdots A_{k+1}$ has $k-1$ minimal circles.

By far, the problem is completely solved. □

Remark. The above three answers are all good with respective peculiarity. As for the one offered by Long Yun, which is concise and throws light on people. Therefore it is somewhat the most excellent.

Exercises 10

1. Find the greatest positive integer n, ensuring the existence of n convex polygons (including triangles) in a plane, of which every two have a common edge but have no common interior.

2. (Chen Ji) If the minimal distance between five points in a plane is 1, and the maximum distance is less than $2\sin 70°$, try proving or negating that the five points are vertexes of a convex pentagon.

3. (Xiong Bin) Suppose that a convex set is covered by two unit circles. Prove that its area is equal or less than $\arccos(-1/3)+2\sqrt{2}$.

Chapter 11 Inequalities for tetrahedron

Triangle is the simplest polygon in plane, while tetrahedron the simplest polyhedron in three-dimensional space. Therefore, the latter can be viewed as the extension of the former in space. In fact, many inequalities for triangle can be extended for tetrahedron. And rich results on the geometric inequalities and the extremal problems of tetrahedron have been attained. Here I would like to introduce several typical examples.

Example 1. If d is the minimal distance between the three pairs of opposite edges of a tetrahedron, and h is the minimal height of the tetrahedron. Show that:

$$2d > h.$$

(A problem of 24th Mathematical Olympiad of Soviet Union)

It is a typical problem in solid geometry, to which the key is to find on a plane on which contains much of numerical relationship. In this way, a problem in space is transformed into a problem in plane.

Proof. Without loss of generality, suppose that $AH = h$, and the distance between AC and BD is d, as Figure 11.1 shows. Draw lines AF and CN perpendicular to BD at points F and N, respectively. Obviously,

$$HF \parallel CN.$$

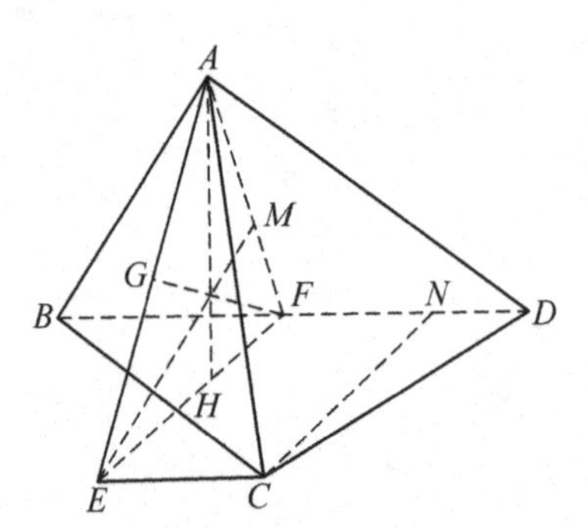

Figure 11.1

Therefore, we can create a rectangle $FECN$ in the plane BCD.

Now let us turn to $\triangle AEF$. AH is the height on side EF, while $FG = d$ is the height on side AE. As for the height EM which equals the distance of the point C to the plane ABD. Therefore, we get: $EM \geqslant AH$.

In this way, all the numerical relationship is showed in the plane AEF, and the problem of solid geometry turns into that of plane geometry. That is to say, we only need to prove that $\frac{AH}{FG} < 2$ in $\triangle AEF$, which is not so difficult.

By $AH \leqslant EM$, we know $AF \leqslant EF$. Besides, $\triangle AEH \backsim \triangle FEG$. Therefore, we get:

$$\frac{h}{d} = \frac{AH}{FG} = \frac{AE}{EF} < \frac{(AF + EF)}{EF} \leqslant 2.$$

□

Remark. The result of Example 1 actually gives out the best estimation of the lower bound for the width of a tetrahedron (the best possible constant is 2). The result was extended to n-dimensional simplex by Ms. Yuan Shufeng and the author. As for the strict estimation of the upper bound for the width of an n-dimensional simplex, the result is given by Mr. Yang Lu and Mr. Zhang Jingzhong. Provided the n-dimensional simplex is specified as a tetrahedron, the result is:

$$d \leqslant \frac{9\sqrt{6}}{2}\sqrt{\sum_{i=1}^{4} \frac{1}{h_i^2}}.$$

The equality holds if and only if the tetrahedron is regular. Therefore, we infer that of all tetrahedrons with the same dimensions and volume the regular tetrahedron has the greatest width. Those who are interested in this problem can refer to the paper by Mr. Yang Lu and Mr. Zhang Jingzhong. (Metric equation and the conjecture of Sallee. Acta Mathematica Sinica (In Chinese, 1983, 26(4):488 - 493.)

The following example mainly makes use of some measuring formulae in tetrahedron.

Example 2. If the radius of tetrahedron $ABCD$'s insphere is r, and $AB = a$, $CD = b$ are a pair of opposite edges. Prove that:

$$r < \frac{ab}{2(a+b)}.$$

(A problem of the 22th Mathematical Olympiad of Soviet Union)

Proof. Suppose that the volume of the tetrahedron is V, and its surface area is S, then

$$r = \frac{3V}{S}. \tag{a}$$

And according to Steiner's theorem, we know

$$V = \frac{1}{6} abd \sin\theta, \tag{b}$$

where d is the distance between the opposite edges AB and CD, and θ is the angle formed by the two edges.

From (a) and (b), we obtain

$$r \leqslant \frac{1}{2} \frac{abd}{S}. \tag{c}$$

On the other hand, as can be shown in Figure 11.2, the distances from A and B to the edge CD are or greater or equal to d. In fact, the distance from one point of the two must be greater than d, therefore:

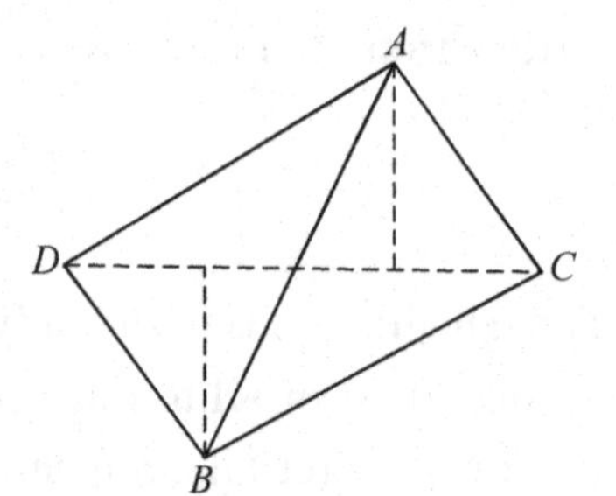

Figure 11.2

$$S_{\triangle ADC} + S_{\triangle BDC} > bd.$$

In like manner, we have

$$S_{\triangle DBA} + S_{\triangle CAB} > ad.$$

Add them up, we have:

$$S > (a+b)d. \tag{d}$$

Therefore, from (c) and (d) we have:

$$r < \frac{abd}{2(a+b)d} = \frac{ab}{2(a+b)}.$$

□

The following example is rather difficult.

Example 3. If r is the radius of tetrahedron $A_1A_2A_3A_4$*'s* insphere, and r_1, r_2, r_3 and r_4 are respectively the radius of the incircles of $\triangle A_2A_3A_4$, $\triangle A_1A_3A_4$, $\triangle A_1A_2A_4$ and $\triangle A_1A_2A_3$, show that:

$$\frac{1}{r_1^2} + \frac{1}{r_2^2} + \frac{1}{r_3^2} + \frac{1}{r_4^2} \leqslant \frac{2}{r^2},$$

and the equality holds if and only if the tetrahedron $A_1A_2A_3A_4$ is regular.

First we need a simple lemma.

Lemma 1. If V is the volume of the tetrahedron $A_1A_2A_3A_4$, and S_1 is the area of $\triangle A_2A_3A_4$ (and so on), and a_{12} is the length of the edge A_3A_4 (and so on), for any $1 \leqslant i \leqslant j \leqslant 4$ we have:

$$\frac{a_{ij}}{S_iS_j \sin\theta_{ij}} = \frac{2}{3V},$$

where θ_{ij} is the dihedral angle formed by the plane opposite to the vertex A_i and the plane opposite to the vertex A_j.

Proof. We need only to confirm that

$$\frac{a_{12}}{S_1S_2 \sin\theta_{12}} = \frac{2}{3V}.$$

As Figure 11. 3 shows, draw a line A_1H, ensure that it is perpendicular to the plane $A_2A_3A_4$ with H as the foot. Then we draw line HD, which is perpendicular to A_3A_4. So we get:

$$3V = A_1H \cdot S_1 = S_1 \cdot A_1D \cdot \sin\theta_{12}$$
$$= S_1 \frac{A_1D \cdot A_3A_4}{a_{12}} \sin\theta_{12} = \frac{2S_1S_2\sin\theta_{12}}{a_{12}}.$$

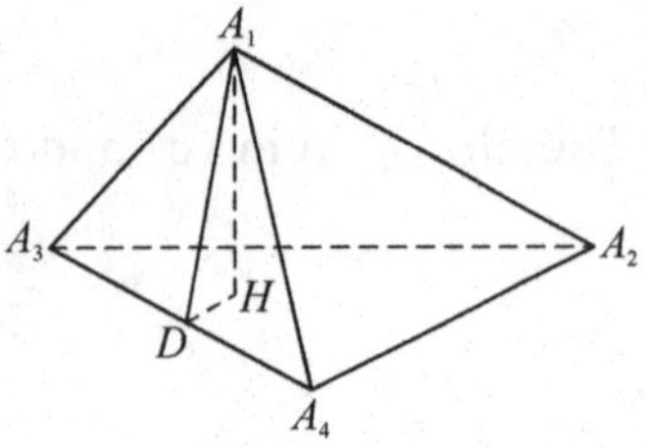

Figure 11.3

The lemma is therefore proved. □

Then let us turn to the original problem.

Proof. If the surface area of $A_1A_2A_3A_4$ is S, according to the Lemma 1 and the formula $3V = rS$, we have:

$$a_{ij} = \frac{2}{rS}S_iS_j\sin\theta_{ij}.$$

So

$$\sum_{j\neq i} a_{ij} = \left(\frac{2}{rS}\right)S_i \sum_{j\neq i} S_j \sin\theta_{ij}. \qquad \text{(a)}$$

Besides, by the projection formula of tetrahedron,

$$S_i = \sum_{j\neq i} S_j \cos\theta_{ij},$$

and by Cauchy's formula, we have

$$\sum_{j\neq i} S_j \sin\theta_{ij} = \sum_{j\neq i} \sqrt{(S_j + S_j\cos\theta_{ij})(S_j - S_j\cos\theta_{ij})}$$
$$\leqslant \left(\sum_{j\neq i} S_j + S_j\cos\theta_{ij}\right)^{\frac{1}{2}} \left(\sum_{j\neq i} S_j - S_j\cos\theta_{ij}\right)^{\frac{1}{2}} \qquad \text{(b)}$$
$$= S^{\frac{1}{2}}(S - 2S_i)^{\frac{1}{2}}.$$

Notice that:

$$\frac{2S_i}{\sum_{j\neq i} a_{ij}} = r_i. \qquad \text{(c)}$$

So according to (a), (b) and (c), we have:

$$\frac{1}{r_i} \leqslant \left(\frac{1}{rS}\right) \cdot S^{\frac{1}{2}}(S - 2S_i)^{\frac{1}{2}}.$$

Therefore

$$\frac{r}{r_i} \leqslant \left(\frac{S - 2S_i}{S}\right)^{\frac{1}{2}}.$$

So

$$\sum_{i=1} \frac{r^2}{r_i^2} \leqslant \sum_{i=1}^{4} \left(\frac{S - 2S_i}{S}\right) = 2.$$

That is

$$\frac{1}{r_1^2} + \frac{1}{r_2^2} + \frac{1}{r_3^2} + \frac{1}{r_4^2} \leqslant \frac{2}{r^2}.$$

□

Remark. The background of this problem is the famous Pölya's inequality for triangle area:

$$S \leqslant \frac{\sqrt{3}}{4}(abc)^{\frac{2}{3}}.$$

During the period from the late 50s to the early 60s of the last century, several scholars extended independently Pölya's inequality to n-dimensional simplex. And the result is now called the volume optimal volume theorem for simplex. Specifically, for a tetrahedron we have

$$V \leqslant \frac{2^{\frac{3}{2}}}{3^{\frac{7}{4}}}\left(\prod_{k=1}^{4} S_k\right)^{\frac{3}{8}}.$$

Naturally, a parallel problem is put forward: is there a similar optimal inequality about the radius of a tetrahedron's insphere and the radiuses of each of its face's incircles?

To answer this question, I wrote the inequality of above example and its deduction $r_1r_2r_3r_4 \geqslant 4r^4$ in a private communication with Canadian mathematician Klamkin in 1992. Later this result was published on Crux. Math. (Problem. 1990, 1994), as recommended by Klamkin. Before long, Mr. Tang Lihua and I extended the result to n-dimensional simplex, and the result was published on Geom.

Dedicata, 1996: 61. Actually, the method used on Geom. Dedicata is different from that on Crux. Math., and the former is completely applicable in n-dimensional space.

It must be pointed out that more profound extension of the optimal volume theorem of simplex is gained by Mr. Zhang Jingzhong and Mr. Yang Lu, and has become the classic result that is widely quoted in the study of distance geometry and geometric inequalities. For more information, consult the paper: Zhang Jingzhong and Yang Lu. On the geometric inequalities of the mass group (in Chinese). The Journal of University of Science and Technology of China, 198111(2): 18.

Still there is another parallel problem about the optimal area theorem, that is, what is the relationship between the radius of a tetrahedron's circumsphere and the radiuses of all its face's circumcircles? The following example provides an answer.

Example 4. If a tetrahedron contains a circumcenter, and the radius of its circumsphere is R, and $R_1R_2R_3$ and R_4 are the radius of $\triangle A_2A_3A_4$, $\triangle A_1A_3A_4$, $\triangle A_1A_2A_4$ and $\triangle A_1A_2A_3$'s circumcircles, respectively, prove that:

$$1 \leqslant \frac{R}{\max(R_1, R_2, R_3, R_4)} \leqslant \frac{3\sqrt{2}}{4}. \tag{a}$$

Proof. Suppose that $\Sigma = A_1A_2A_3A_4$ is a tetrahedron. O is the circumcenter of Σ, and it is within the body. Obviously, the projection O_i of O on a certain plane must be the circumcenter of the plane's triangle. By $A_iO_i \leqslant A_iO$, $i = 1, 2, 3, 4$, we know $R_i \leqslant R$. And the equality may hold. Therefore, the left side of the inequality in (a) is demonstrated.

Now make a maximum sphere contained in Σ, with O as its centre and d as its radius. It can be inferred that this sphere is at least tangential to certain face of Σ. Without loss of generality, we may suppose it is tangential to the face $A_2A_3A_4$, then the point of tangency must be the circumcenter of $\triangle A_2A_3A_4$, therefore

$$R_1^2 = R^2 - d^2. \qquad \text{(b)}$$

And of all the spheres contained in Σ, its insphere's radius is the greatest. So from the well known inequality $r \leqslant \frac{R}{3}$, we get

$$d \leqslant \frac{R}{3}. \qquad \text{(c)}$$

Then according to (a), (b) and (c), we know

$$R^2 - R_1^2 \leqslant \frac{R^2}{9}.$$

That is,

$$R \leqslant \frac{3}{\sqrt{8}} R_1 \leqslant \frac{3}{\sqrt{8}} \max(R_1, R_2, R_3, R_4).$$

So the right side of the inequality (a) is also proved. □

Example 5. Suppose G is the barycenter of the tetrahedron $A_1A_2A_3A_4$. The distance from G to the edge A_iA_j is h_{ij}, and that from G to A_i is D_i, Prove that:

$$\sum_{1 \leqslant i < j \leqslant 4} h_{ij} \leqslant \frac{\sqrt{3}}{2} \sum_{i=1}^{4} D_i. \qquad \text{(a)}$$

The basic idea is to transform the hight h_{ij} of $\triangle A_iA_jG$ to the correspondent internal angle bisector in a triangle. To achieve that, firstly we need to prove the following lemma.

Lemma 2. If AT is the internal angle bisector of $\triangle ABC$, it can be inferred that

$$AT^2 \leqslant \frac{1}{2}(\overrightarrow{AB} \cdot \overrightarrow{AC} + |\overrightarrow{AB}| \cdot |\overrightarrow{AC}|).$$

Proof. From the angle bisector formula, we know

$$AT = \frac{bc}{2(b+c)}\cos\frac{A}{2} \leqslant \sqrt{bc}\cos\frac{A}{2}.$$

Therefore

$$AT^2 \leqslant bc\cos^2\frac{A}{2} = \frac{1}{2}bc(1+\cos A)$$
$$= \frac{1}{2}(\overrightarrow{AB}\cdot\overrightarrow{AC} + |\overrightarrow{AB}|\cdot|\overrightarrow{AC}|).$$

So the lemma is confirmed. □

Then let us turn to the original problem.

Proof. To make things easier, we have $\overrightarrow{GA_i} = \overrightarrow{A_i}$, $i = 1, 2, 3, 4$. Then

$$\sum_{i=1}^{4}\overrightarrow{A_i} = \vec{0},$$

therefore

$$\left(\sum_{i=1}^{4}\overrightarrow{A_i}\right)^2 = 0.$$

That is

$$\sum_{i=1}^{4}D_i^2 + 2\sum_{1\leqslant i<j\leqslant 4}\overrightarrow{A_i}\cdot\overrightarrow{A_j} = 0. \tag{b}$$

On the other hand, from Lemma 2 we know

$$h_{ij}^2 \leqslant \frac{1}{2}(\overrightarrow{A_i}\cdot\overrightarrow{A_j} + |\overrightarrow{A_i}|\cdot|\overrightarrow{A_j}|),$$

so

$$\left(\sum_{1\leqslant i<j\leqslant 4}h_{ij}\right)^2 \leqslant 6\sum_{1\leqslant i<j\leqslant 4}h_{ij}^2$$
$$\leqslant 3\sum_{1\leqslant i<j\leqslant 4}(\overrightarrow{A_i}\cdot\overrightarrow{A_j} + |\overrightarrow{A_i}|\cdot|\overrightarrow{A_j}|) \tag{c}$$
$$= 3\sum_{1\leqslant i<j\leqslant 4}\overrightarrow{A_i}\cdot\overrightarrow{A_j} + 3\sum_{1\leqslant i<j\leqslant 4}D_i\cdot D_j.$$

By substitution of (b) into (c), we get

$$\begin{aligned}
\left(\sum_{1\leqslant i<j\leqslant 4} h_{ij}\right)^2 &\leqslant -\frac{3}{2}\sum_{i=1}^{4} D_i^2 + 3\sum_{1\leqslant i<j\leqslant 4} D_i \cdot D_j \\
&= \frac{3}{2}\left(-\sum_{i=1}^{4} D_i^2 + \left[\left(\sum_{i=1}^{4} D_i\right)^2 - \sum_{i=1}^{4} D_i^2\right]\right) \\
&= \frac{3}{2}\left[\left(\sum_{i=1}^{4} D_i\right)^2 - 2\sum_{i=1}^{4} D_i^2\right] \qquad \text{(d)} \\
&\leqslant \frac{3}{2}\left[\left(\sum_{i=1}^{4} D_i\right)^2 - \frac{1}{2}\left(\sum_{i=1}^{4} D_i\right)^2\right] \\
&= \frac{3}{4}\left(\sum_{i=1}^{4} D_i\right)^2.
\end{aligned}$$

(The last inequality makes use of Cauchy's inequality.)

By (d), we can easily get the required inequality. □

Remark. The above example is a particular case of Mr. Chen Ji's conjecture, which is:

$$\sum_{1\leqslant i<j\leqslant 4} h_{ij} \leqslant \frac{\sqrt{3}}{2}\sum_{i=1}^{4} R_i,$$

where P is a point within the tetrahedron $A_1A_2A_3A_4$, and h_{ij} is the distance from P to the edge A_iA_j, while R_i is that from P to A_i.

As far as the author knows, the above conjecture has not been solved yet.

Exercises 11

1. If P and Q are two points within the regular tetrahedron $ABCD$, prove that:

$$\angle PAQ < 60°.$$

2. If the distance between the three pairs of opposite edges of a tetrahedron are d_1, d_2, and d_3, respectively. Prove that the volume of the tetrahedron $V \geqslant \frac{1}{3}d_1d_2d_3$. (This is a problem of the 28th

Mathematical Olympiad in Moscow.)

3. If we connect two points on a sphere with radius 1 by a curve whose length is less than 2, prove that the curve must be contained in a hemisphere. (A problem of the 3rd Mathematical Olympiad of USA.)

4. Suppose that I is the incenter of the tetrahedron $A_1A_2A_3A_4$. The area of $\triangle A_iIA_j$ is I_{ij}, and the area of the triangle opposite to A_i is S_i. Try to prove:

$$\sum_{1 \leqslant i < j \leqslant 4} I_{ij} \leqslant \frac{\sqrt{6}}{4} \sum_{i=1}^{4} S_i.$$

5. Suppose R and r are the radius of the circumsphere and the insphere of a tetrahedron, respectively. And O and I are the center of that two spheres, respectively. Prove that:

$$R^2 \geqslant 9r^2 + OI^2.$$

Answers and hints to selected exercises

Chapter 1 The method of segment replacement for distance inequalities

1. Extend BA to C', such that $AC' = AC$, and connect $A'C'$, then $A'B + A'C > BC' = BA + AC$. And $A'C' = A'C$ by $\triangle AA'C' \cong \triangle AA'C$.

2. Let A_1 and A_2, B_1 and B_2, C_1 and C_2 be points on sides BC, CA, AB, respectively, such that B_1C_2, C_1A_2, A_1B_2 pass through point O and $B_1C_2 \parallel BC$, $C_1A_2 \parallel CA$, $A_1B_2 \parallel AB$. The largest sides of $\triangle A_1A_2O$, $\triangle B_1B_2O$, $\triangle C_1C_2O$ are A_1A_2, B_1O, C_2O, respectively. So $OP < A_1A_2$, $OQ < B_1O$, $OR < C_2O$, then $OP + OQ + OR < A_1A_2 + B_1O + C_2O = A_1A_2 + CA_2 + BA_1 = BC$.

3. *Hint*. Without loss of generality, let $A \leqslant 90°$. Reflect $\triangle ABC$ on line AC to $\triangle AB'C$, and then reflect $\triangle AB'C$ on line AB' to $\triangle AB'C'$. Simultaneously, $\triangle DEF$ turns to $\triangle D'EF'$ and $\triangle D''E''F'$ in turn, then $DE + EF + FD = DE + EF' + F'D'' \geqslant DD''$. Then to prove $DD'' \geqslant \frac{2\Delta}{R}$.

4. We only give hints for points E, F on AC, AB, respectively, for other cases can obtain similarly. Without loss of generality, suppose that $AC \geqslant AB$, and take point D on AC such that $AD = AB$; take point G in AB such that $AG = AE$, then

$$\begin{aligned} &|AB - AC| + |AE - AF| = |CD| + |FG| \\ \geqslant &|CG - DG| + |CF - CG| \\ \geqslant &|CG - DG + CF - CG| \\ = &|CF - DG| = |BE - CF|. \end{aligned}$$

5. Construct rays PX, PY, such that $\angle XPY = \frac{\pi}{3}$. Points B_1, B_3, B_5 and B_2, B_4, B_6 are on PX, PY respectively, such that $PB_i = OA_i$, then $B_iB_{i+1} = A_iA_{i+1}$. Because $PB_1 > PB_3 > PB_5$, $PB_6 < PB_4 < PB_2$, so segment B_1B_6 must intersect B_2B_3, B_3B_4, B_4B_5 at any points C_1, C_2, C_3. Because of $B_1B_2 < B_1C_1 + C_1B_2$, $B_3C_2 < B_3C_1 + C_1C_2$, $C_2B_4 < C_2C_3 + C_3B_4$, $B_5B_6 < B_5C_3 + C_3B_6$, adding up these results, yields $B_1B_2 + B_3B_4 + B_5B_6 < B_2B_3 + B_4B_5 + B_6B_1$.

Chapter 2 Ptolemy's inequality and its application

1. Let $AC = a$, $CE = b$ and $AE = c$, applying Ptolemy's inequality to quadrilateral $ACEF$, $AC \cdot EF + CE \cdot AF \geqslant AE \cdot CF$. Since $EF = AF$, so $FA/FC \geqslant c/(a+b)$. Similarly, $DE/DA \geqslant b/(c+a)$, $BC/BE \geqslant a/(b+c)$. So $BC/BE + DE/DA + FA/FC \geqslant a/(b+c) + b/(c+a) + c/(a+b) \geqslant 3/2$, the equality holds if $ABCDEF$ is an inscribed hexagon and $a = b = c$.

2. Let $ABDE$ be a square, then $OM = \frac{CE}{2}$, $ON = \frac{CD}{2}$, so $OM + ON = (CD + CE)/2$. Let $AB = c$, then $BD = AE = c$, $AD = BE = \sqrt{2}c$. Applying generalized Ptolemy's theorem to quadrilateral $ACBD$ and $ACBE$, $OM + ON = (CD + CE)/2 \leqslant (\sqrt{2}+1)(a+b)/2$. So the maximum of $OM + ON$ is $(\sqrt{2}+1)(a+b)/2$.

3. Connect BD, AE, since $AB = BC = CD$, $\angle BCD = 60°$, so $BD = AB$. Similarly, $AE = ED$, so points A and D are symmetrical to line BE. Construct symmetrical points C', F' of C, F to line BE. Then $\triangle ABC'$ and $\triangle DEF'$ are equilateral triangles, points G and H are in the circumcircle of these two triangles. Applying Ptolemy's theorem to quadrilateral $AGBC'$ and $DHEF'$ and noticing that segments CF and $C'F'$ are symmetrical to line BE, we can get the conclusion immediately.

4. Construct the circumcircle O_1 of $\triangle AGH$, which intersects AD

at Q. It is easy to see $\triangle BCD \backsim \triangle APE$, so $\frac{DC}{PE} = \frac{BC}{AP} = \frac{BD}{AE}$, that is $DC = \frac{PE}{AP} \cdot BC = \frac{AK}{AP} \cdot BC$, $BD = \frac{AE}{AP} \cdot BC$. Applying Ptolemy's theorem to quadrilateral $ABDC$, $AD \cdot BC = BD \cdot AC + DC \cdot AB = \frac{AE}{AP} \cdot BC \cdot AC + \frac{AK}{AP} \cdot BC \cdot AB$, so $AP \cdot AD = AE \cdot AC + AK \cdot AB$……(1). Similarly, applying Ptolemy's theorem, we get $AP \cdot AQ = AE \cdot AH + AK \cdot AG$. So $AP^2 + PG \cdot PH = AP^2 + AP \cdot PQ = AE \cdot AH + AK \cdot AG$, then $AP^2 = AE \cdot AH + AK \cdot AG - PG \cdot PH$……(2). (1) − (2), $AP(AD - AP) = AE(AC - AH) + AK(AB - AG) + PG \cdot PH$, that is $PA \cdot PD = PK \cdot PI + PE \cdot PF + PG \cdot PH$. Also

$$PK \cdot PI \leqslant \left(\frac{PK + PI}{2}\right)^2 = \frac{1}{4}KI^2,$$

$$PE \cdot PF \leqslant \frac{1}{4}EF^2, \qquad PG \cdot PH \leqslant \frac{1}{4}GH^2,$$

so $EF^2 + KI^2 + GH^2 \geqslant 4PA \cdot PD$, the equality holds if and only if P is the barycenter of $\triangle ABC$.

5. Applying Ptolemy's theorem to quadrilateral ACA_1B, $AA_1 \cdot BC = AB \cdot A_1C + AC \cdot A_1B$. Let $A_1B = A_1C = x$, notice that $2x = A_1B + A_1C > BC$, we have $2AA_1 = 2(ABx + ACx)/BC = (AB + AC) \cdot 2x/BC > AB + AC$. That is $AA_1 > (AB + AC)/2$. Similarly, $BB_1 > (BA + BC)/2$, and $CC_1 > (CA + CB)/2$, adding up these three formulas, the conclusion is clear.

Chapter 3 Inequality for the inscribed quadrilateral

1. *Solution.* Using the cosine law, we have

$$BD^2 = AD^2 + AB^2 - 2AD \cdot AB\cos A = CD^2 + BC^2 - 2CD \cdot BC\cos C,$$

since $60° \leqslant A \leqslant 120°$, $60° \leqslant C \leqslant 120°$, it follows that

$$-\frac{1}{2} \leqslant \cos A \leqslant \frac{1}{2}, \qquad -\frac{1}{2} \leqslant \cos C \leqslant \frac{1}{2},$$

therefore

$$\begin{aligned}&3BD^2-(AB^2+AD^2+AB\cdot AD)\\&=2(AD^2+AB^2)-AD\cdot AB(1+6\cos A)\\&\geqslant 2(AD^2+AB^2)-4AD\cdot AB\\&=2(AB-AD)^2\geqslant 0,\end{aligned}$$

thus

$$\begin{aligned}&\frac{1}{3}(AB^2+AD^2+AB\cdot AD)\\&\leqslant BD^2\\&=CD^2+BC^2-2CD\cdot BC\cos C\\&=CD^2+BC^2+CD\cdot BC.\end{aligned}$$

Since $ABCD$ has an incircle, we have

$$AD+BC=AB+CD, \tag{1}$$

therefore

$$|AB-AD|=|CD-BC|. \tag{2}$$

Concluding from these relations, we get

$$\frac{1}{3}|AB^3-AD^3|\leqslant|BC^3-CD^3|,$$

and the equality holds if $\cos A=1/2$, $AB=AD$, $\cos C=-1/2$ or $|AB-AD|=|CD-BC|=0$. Thus the equality holds, when $AB=AD$ or $CD=BC$. We can prove the other inequality in the same way.

2. *Answer*. For the convex quadrilateral $ABCD$, let the circumcenter be O, and set $EF\perp MN$, $F\in MN$. Since points B, M, E, N are on the same circle with diameter BE, we have

$$MN=BE\cdot\sin B=\frac{BE\cdot AC}{2R}$$

and

$$EF=EM\cdot\sin\angle EMN=\frac{AE\cdot BE\sin\angle AEB}{AB}\cdot\sin\angle CEB,$$

and combining

$$BE \cdot \sin\angle EBC = CE \cdot \sin\angle BEC = \frac{CE \cdot BE}{2R}$$

with

$$AE \cdot EC = R^2 - OE^2,$$

we obtain

$$EF = \frac{R^2 - OE^2}{2R} \cdot \sin\angle AEB,$$

therefore

$$S_{\triangle MEN} = \frac{1}{2}MN \cdot EF = \frac{(R^2 - OE^2) \cdot AC \cdot BE \cdot \sin\angle AEB}{8R^2},$$

similarly, we have $S_{\triangle MEP}$, $S_{\triangle PEQ}$, $S_{\triangle QEM}$, then summing up, yields

$$S_{MNPQ} = \frac{(R^2 - OE^2) \cdot AC \cdot BD \cdot \sin\angle AEB}{4R^2} = \frac{(R^2 - OE^2) \cdot S}{2R^2} \leqslant \frac{S}{2}.$$

3. *Proof*. Since of all quadrilateral with given sides, the one who inscribed in a circle has the largest area, we need only consider the case that two convex quadrilaterals both have circumcircle.

In this case, $S = \sqrt{(s-a)(s-b)(s-c)(s-d)}$, with $s = (a + b + c + d)/2 = a + c = b + d$. Likewise for S'.

Using the A-G inequality, we obtain

$$\begin{aligned}
& aa' + bb' + cc' + dd' \\
= {} & (s-a)(s'-a') + (s-b)(s'-b') + (s-c)(s'-c') + \\
& (s-d)(s'-d') \\
\geqslant {} & 4[(s-a)(s'-a')(s-b)(s'-b')(s-c)(s'-c') \times \\
& (s-d)(s'-d')]^{\frac{1}{4}} \\
= {} & 4\sqrt{SS'}
\end{aligned}$$

as desired.

4. *Omitted*.

5. *Proof*. Assume $AB = a$, $BC = b$, $CD = c$, $DA = d$, so we have

$$a = \rho_a\left(\tan\frac{A}{2} + \tan\frac{B}{2}\right),$$

therefore,

$$\frac{a}{\rho_a} \geqslant \sqrt{\tan\frac{A}{2}\tan\frac{B}{2}},$$

likewise

$$\frac{b}{\rho_b} \geqslant \sqrt{\tan\frac{C}{2}\tan\frac{B}{2}},$$

$$\frac{c}{\rho_c} \geqslant \sqrt{\tan\frac{C}{2}\tan\frac{D}{2}},$$

$$\frac{d}{\rho_d} \geqslant \sqrt{\tan\frac{D}{2}\tan\frac{A}{2}}.$$

Since $A + B = C + D = \pi$, we obtain

$$\tan\frac{A}{2}\tan\frac{C}{2} = \tan\frac{B}{2}\tan\frac{D}{2} = 1,$$

therefore

$$\begin{aligned}\frac{1}{\rho_a} + \frac{1}{\rho_c} &\geqslant \frac{2}{a}\sqrt{\tan\frac{D}{2}\tan\frac{A}{2}} + \frac{2}{c}\sqrt{\tan\frac{D}{2}\tan\frac{A}{2}} \\ &\geqslant 2\sqrt{\frac{4}{ac}\sqrt{\tan\frac{A}{2}\tan\frac{B}{2}\tan\frac{C}{2}\tan\frac{D}{2}}} \\ &= \frac{4}{\sqrt{ac}},\end{aligned}$$

similarly

$$\frac{1}{\rho_b} + \frac{1}{\rho_d} \geqslant \frac{4}{\sqrt{bd}},$$

thus

$$\frac{1}{\rho_a} + \frac{1}{\rho_b} + \frac{1}{\rho_c} + \frac{1}{\rho_d} \geqslant \frac{4}{\sqrt{ac}} + \frac{4}{\sqrt{bd}} \geqslant \frac{8}{\sqrt[4]{abcd}},$$

the equalities hold if and only if $A = B = C = D$, and $a = b = c = d$.

Chapter 4 The area inequality for special polygons

1. *Answer*. Without loss of generality, assume $C > 90°$, so $\min\{\angle A, \angle B\} < 45°$. Without loss of generality assume $\angle A < 45°$. Viewing AB as the diameter, draw the semicircle O in the same side as the vertex C, so C located within semicircle. Construct radial AT such that $\angle BAT = 45°$, then construct radial OE such that $\angle BOE = 45°$, and the intersection points is E with the semicircle. Draw a semicircle tangent through E, intersect with the extension of AT, and AT at the points D and E, respectively, then isosceles right triangle ADF covers the ABC, and

$$\begin{aligned} AD &= AO + OD = \frac{1}{2}AB + \frac{\sqrt{2}}{2}AB \\ &= \frac{1}{2}(1+\sqrt{2}) \cdot AB \\ &\leqslant \frac{1}{2}(1+\sqrt{2}) \cdot 2R = 1+\sqrt{2}, \end{aligned}$$

as desired.

2. *The answer is similar to Example* 2.

3. *Omitted*.

4. *Hint*. In a square with side length one, a regular triangle with length more than $\frac{\sqrt{2}}{3}$ must contain the centre of the square.

5. *Answer*. Suppose that any three points of given five points A_1, A_2, A_3, A_4, A_5 on plane are not co-linear.

(1) If the convex hull of the five points is not a convex pentagon, then there is one point locating in some triangle. It is easy to prove $\mu_5 \geqslant 3$.

(2) If the convex hull of the five points is a convex pentagon $A_1A_2A_3A_4A_5$. Draw $MN \parallel A_3A_4$ and intersect with A_1A_3, A_1A_4, at points M, N, respectively, such that

$$\frac{A_1M}{MA_3}=\frac{A_1N}{NA_4}=\frac{\sqrt{5}-1}{2}.$$

(i) For A_2, A_5, if A_2, A_3 and A_4 lie in the same side of MN, it follows that

$$\mu_5 \geqslant \frac{S_{\triangle A_1A_3A_4}}{S_{\triangle A_2A_3A_4}} \geqslant \frac{A_1A_3}{MA_3}=1+\frac{A_1M}{MA_3}=\frac{\sqrt{5}+1}{2}.$$

(ii) If A_2, A_5 and A_1 lie on the same side of MN, then let A_2A_5 meet A_1A_3 at point O, so $A_1O \leqslant A_1M$, thus

$$\mu_5 \geqslant \frac{S_{\triangle A_2A_3A_5}}{S_{\triangle A_1A_2A_5}}=\frac{OA_3}{OA_1} \geqslant \frac{MA_3}{MA_1}=\frac{\sqrt{5}+1}{2}.$$

Notice that $3 \geqslant (\sqrt{5}+1)/2$, then we have $\mu_5 \geqslant (\sqrt{5}+1)/2$.

If A_1, A_2, A_3, A_4, A_5 are the vertexes of regular pentagon which the side length a, then we obtain

$$\mu_5=\frac{S_{\triangle A_1A_3A_4}}{S_{\triangle A_1A_2A_3}}=\frac{\frac{1}{2}A_1A_3 \cdot A_1A_4}{\frac{1}{2}A_1A_2 \cdot A_1A_3}=\frac{A_1A_4}{A_1A_2}=\frac{\sqrt{5}+1}{2}.$$

From that we know the minimal value of μ_5 is $(\sqrt{5}+1)/2$.

Chapter 5 Linear geometric inequalities

1. The problem can be turned to prove

$$AA_1+BB_1+CC_1 \geqslant \frac{4}{3}(m_a+m_b+m_c),$$

where m_a is the mid-line of side a, and so on. Let $AA_1=M_a$, and so on. By chord intersection theorem, we have $m_a \cdot (M_a-m_a)=a^2/4$, combine with $4m_a^2=8k^2-3a^2$, we have $3M_a=2k(k/m_a+m_a/k) \geqslant 4k$ and so on, where $k^2=(a^2+b^2+c^2)/4$, $(k>0)$.

2. Suppose M, N, P, Q are four given points in convex quadrangle $ABCD$. It suffices to consider the case of $MNPQ$ is a convex quadrangle. Let E be any boundary point of $ABCD$. Let $f(E)=$

$EA + EB + EC + ED$ and $g(E) = EM + EN + EP + EQ$, thus it suffices to prove that there is a boundary point E such that $f(E) > g(E)$. Make line MP intersecting quadrangle $ABCD$ at F, G, then $f(F) + f(G) > g(F) + g(G)$, which shows $f(F) > g(F)$ or $f(G) > g(G)$.

3. *Omitted*.

4. Without loss of generality, suppose that $AE \leqslant AC \leqslant CE$. By Ptolemy's theorem, we have $AD \cdot CE \leqslant AC \cdot DE + AE \cdot CD \leqslant a(AC + AE) \leqslant 2a \cdot CE$, thus $AD \leqslant 2a$.

5. We need to use the following trigonometric inequalities

$$\cos(\beta + \gamma)\cos\alpha + \cos(\gamma + \alpha)\cos\beta + \cos(\alpha + \beta)\cos\gamma \leqslant \frac{3\sqrt{6} + 9\sqrt{2}}{8},$$

$$\tan\alpha \cdot \tan\beta + \tan\beta \cdot \tan\gamma + \tan\gamma \cdot \tan\alpha \leqslant 3(7 - 4\sqrt{3}),$$

$$\sin\alpha\sin\beta\sin\gamma \leqslant \frac{3\sqrt{6} - 5\sqrt{2}}{16},$$

where $A = 4\alpha$, $B = 4\beta$, $C = 4\gamma$.

Chapter 6 Algebraic methods

1. *Hint*. Using complex number identities or Ptolemy's theorem.

2. Without loss of generality, let $A = (0, h)$, $B = (p, 0)$, $C = (q, 0)$ $(h > 0, p < q)$, thus $E = (-h, h - p)$, $G = (h, h + q)$. The equation of line EG is

$$y = h + \frac{q - p}{2} + \frac{p + q}{2h}x.$$

Let $x = p$ and $x = q$, we have

$$BP = h + \frac{q - p}{2} + \frac{p + q}{2h}p, \quad CQ = h + \frac{q - p}{2} + \frac{p + q}{2h}q.$$

Notice that

$$EG^2 = 4h^2 + (p + q)^2 \geqslant 4h^2,$$

namely, $EG \geqslant 2h$, hence

$$BP + CQ = q - p + \frac{4h^2 + (p + q)^2}{2h} = BC + \frac{EG^2}{2h} \geqslant BC + EG.$$

3. *Omitted*.

4. Let $p = (a + b + c)/2$, $x = p - a$, $y = p - b$, $z = p - c$, thus $x, y, z > 0$. Let the area of $\triangle ABC$ be S, thus $2S = ah_a = 2(p - b)r_b$, we have

$$\frac{h_a}{r_b} = \frac{2(p - b)}{a} = \frac{2y}{y + z},$$

then

$$\sum \left(\frac{h_a}{r_b}\right)^2 = 4 \sum \frac{y^2}{(y + z)^2}.$$

Since

$$\sum \sin^2 \frac{A}{2} = \sum \frac{(p - b)(p - c)}{bc} = \sum \frac{yz}{(x + y)(z + x)},$$

hence the original inequality is equivalent to

$$\sum \frac{y^2}{(y + z)^2} \geqslant \sum \frac{yz}{(x + y)(z + x)}$$

$$\Leftrightarrow \sum y^2 (x + y)^2 (z + x)^2 \geqslant \sum yz(x + y)(z + x)(y + z)^2$$

$$\Leftrightarrow \sum y^2 \left(yz + x \sum x\right)^2 \geqslant \sum yz \left(yz + x \sum x\right)(y + z)^2$$

$$\Leftrightarrow \sum y^2 x^2 \left(\sum x\right)^2 + \sum y^4 z^2 + 2xyz \sum x \sum y^2$$

$$\geqslant \sum y^2 z^2 (y + z)^2 + xyz \sum (y + z)^2 \sum x$$

$$\Leftrightarrow \sum y^2 z^2 (x^2 + 2xy + 2xz) + \sum y^4 z^2 \geqslant 2xyz \sum x \sum xy$$

$$= 6x^2 y^2 z^2 + 2xyz \left(\sum x^2 (y + z)\right)$$

$$\Leftrightarrow 3x^2 y^2 z^2 + 2xyz \sum x^2 (y + z) \leqslant 2xyz \sum yz(y + z) + \sum y^4 z^2$$

$$\Leftrightarrow 3x^2 y^2 z^2 \leqslant \sum y^4 z^2.$$

By mean value inequality, the last inequality follows.

5. (1) Prove that the boundary Γ of locus P is an ellipse by analytical method. (2) Create a new Cartesian coordinate system. Let the equation of Γ be Γ: $\frac{x^2}{a^2}+\frac{y^2}{b^2}=1$ $(a, b>0)$, thus $\triangle ABC$ is circumscribed triangle of Γ, $\triangle DEF$ is inscribed triangle of Γ. By transformation $x=ax'$, $y=by'$, so that an arbitrary convex region D in xOy plane is transformed to an convex region D' in $x'Oy'$ plane. Γ is transformed to a unit circle. There is an area relation $|D|=ab\cdot|D'|$. Let $\triangle ABC$, $\triangle DEF$ be transformed to $\triangle A'B'C'$, $\triangle D'E'F'$ respectively, thus $\triangle A'B'C'$, $\triangle D'E'F'$ are circumscribed triangle and circumscribed triangle of $\odot O'$, respectively. We have $S_\Gamma=ab\cdot S_{\odot O'}$, $S_{\triangle ABC}=ab\cdot S_{\triangle A'B'C'}$, $S_{\triangle DEF}=ab\cdot S_{\triangle D'E'F'}$, hence

$$\frac{S_\Gamma}{S_{\triangle DEF}}=\frac{S_{\odot O'}}{S_{\triangle D'E'F'}}\geqslant\frac{4\sqrt{3}\pi}{9},\qquad \frac{S_\Gamma}{S_{\triangle ABC}}=\frac{S_{\odot O'}}{S_{\triangle A'B'C'}}\leqslant\frac{\sqrt{3}}{9}\pi.$$

Chapter 7 Isoperimetric and extremal value problem

1. (1) Let the vertex A locate in circular arc with BC its chord and $\angle A=\alpha$. If point A is the farthest to line BC. Namely if A locate on vertical bisector of $\triangle ABC$, then the area of $\triangle ABC$ is maximum.

(2) For given side BC and angle α, the radius of circumcircle of $\triangle ABC$ is constant. So $AB+AC=2R(\sin\gamma+\sin\beta)=4R\cdot\sin((\pi-\alpha)/2)\cos((\gamma-\beta)/2)$. Thus, if and only if $\cos((\gamma-\beta)/2)=1$, namely, $\gamma=\beta$, then $AB+AC$ is maximal.

2. Let $\triangle ABC$ and $\triangle PQR$ be regular, D and E be intersection points of AB and AC with PQ, respectively. By the symmetry of rotation, we get $K=S_{\triangle ABC}-3S_{\triangle ADE}=\frac{3\sqrt{3}}{4}r^2-3S_{\triangle ADE}$. Notice that circumference of $\triangle ADE$ is a constant $\sqrt{3}r$. By isoperimetric theorem, the maximal area of $\triangle ADE$ is attained if $\triangle ADE$ is regular. So $K\geqslant\frac{\sqrt{3}}{2}r^2$.

3. Let the three sides of a quadrilateral $ABCD$ satisfy $AB=BC=$

$CD = 1$. It is clear that if a quadrilateral has the largest area it must be convex. By symmetry, we need only to consider the cases that $\angle A = 30°$ or $\angle B = 30°$.

(1) In the case of $\angle B = 30°$, it is easy to see that the largest area of quadrilateral $ABCD$ is

$$S_1 = \frac{\sqrt{6} - \sqrt{2} + 1}{4}.$$

(2) In the case of $\angle A = 30°$, make symmetric points B', C' of B, C to the line AD. Joint AB', $B'C'$, $C'D$, BB', then $\triangle ABB'$ is a regular triangle with sides 1. The side lengths of pentagon $BCDC'D'$ are all 1. The pentagon has the maximum area $(\sqrt{25+10\sqrt{5}})/4$ if and only if it is regular. So maximal area of quadrilateral $ABCD$ is

$$S_2 = \frac{1}{2}\left[\frac{\sqrt{3}}{4} + \frac{\sqrt{25+10\sqrt{5}}}{4}\right] = \frac{\sqrt{3}}{8} + \frac{\sqrt{25+10\sqrt{5}}}{8}.$$

It is easy to see $S_2 > S_1$, $S_{\max} = \frac{\sqrt{3}}{8} + \frac{\sqrt{25+10\sqrt{5}}}{8}$ and in the quadrilateral $ABCD$, $\angle A = 30°$, $\angle C = 108°$, $\angle D = 54°$, $AB = BC = CD = 1$.

4. Let the area of n-gon $A_1A_2\cdots A_n$ be S. By the area relation, we have $\sum_{i=1}^{n} a_i d_i = 2S$. By Cauchy inequality and isoperimetric inequality, we obtain

$$\sum_{i=1}^{n} \frac{a_i}{d_i} = \sum_{i=1}^{n} \frac{a_i^2}{a_i d_i} \geqslant \frac{\left(\sum a_i\right)^2}{\sum_{i=1}^{n} a_i d_i} = \frac{\left(\sum a_i\right)^2}{2S}$$

$$\geqslant \frac{1}{2S} \cdot 4n \cdot S \cdot \tan\frac{\pi}{n} = 2n\tan\frac{\pi}{n}.$$

5. *Hint*. Using the same method similar to Example 3.

Chapter 8 Embed inequality and inequality for moment of inertia

1. *Hint*. Substitute $\cos A = (b^2 + c^2 - a^2)/(2bc)$ into the embedding triangle inequality and simplify.

2. Notice the formula $\cot A = (b^2 + c^2 - a^2)/(4\Delta)$, etc., the original inequality is equivalent to $a^2 + b^2 + c^2 \geqslant 4\Delta\cos A'/\sin A + 4\Delta\cos B'/\sin B + 4\Delta\cos C'/\sin C$. Then, by the area formula, the previous formula is equivalent to $a^2 + b^2 + c^2 \geqslant 2ab\cos C' + 2ac\cos B' + 2bc\cos A'$. This is a special case of the embedding inequality.

3. *Hint*. In the embedding inequality, let $x = \tan(A/2)$, $y = \tan(B/2)$ and $z = \tan(C/2)$. By trigonometric identities, we can obtain the results.

4. Let $u = x\sin\alpha + y\sin\beta$, $v = z\sin\gamma + w\sin\theta$, then $u^2 = (x\sin\alpha + y\sin\beta)^2 \leqslant (x\sin\alpha + y\sin\beta)^2 + (x\cos\alpha - y\cos\beta)^2 = x^2 + y^2 - 2xy\cos(\alpha + \beta)$. Similarly, we can get $v^2 \leqslant z^2 + w^2 - 2zw\cos(\gamma + \theta)$. Taking notice of the known conditions $\alpha + \beta + \gamma + \theta = (2k + 1)\pi$, $(k \in \mathbf{Z})$, we will get $\cos(\alpha + \beta) + \cos(\gamma + \theta) = 0$, so $\dfrac{x^2 + y^2 - u^2}{2xy} + \dfrac{z^2 + w^2 - v^2}{2zw} \geqslant 0$, that is $u^2/(xy) + v^2/(zw) \leqslant (x^2 + y^2)/(xy) + (z^2 + w^2)/(zw) = (xz + yw)(xw + yz)/(xyzw)$. In addition, applying Cauchy inequality, we have $|u + v| \leqslant \sqrt{(xy + zw)(xz + yw)(xw + yz)/(xyzw)}$ as required.

5. In inequality for moment of inertia, let $x = a_2^2 + a_3^2 - a_1^2$ and so on, we have the left $\geqslant \sum(a_3^2 + a_1^2 - a_2^2)(a_1^2 + a_2^2 - a_3^2) \cdot a_1^2/(a_1^2 + a_2^2 + a_3^2) \cdots (1)$. Notice also that $a_1^2 + a_2^2 + a^2 + a_3^2 \leqslant 9R^2$, $R = a_1a_2a_3/(4\Delta)$. Combining the previous inequalities, we have $a_1^2a_2^2a_3^2/(a_1^2 + a_2^2 + a_3^2) \geqslant 16\Delta^2/9 \cdots (2)$. Further by Heron formula $16\Delta^2 = 2(a_1^2a_2^2 + a_2^2a_3^2 + a_3^2a_1^2) - (a_1^4 + a_2^4 + a_3^4) \cdots (3)$. Thus, by (1), (2), and (3), we have $\sum(a_3^2 + a_1^2 - a_2^2)(a_1^2 + a_2^2 - a_3^2) \cdot a_1^2/(a_1^2 + a_2^2 + a_3^2) \geqslant -16\Delta^2 + 12a_1^2a_2^2a_3^2/(a_1^2 + a_2^2 + a_3^2) \geqslant -16\Delta^2 + 12 \cdot 16\Delta^2/9 = 16\Delta^2/3$.

Chapter 9 Locus problem of Tsintsifas's inequality

1. Applying Gergonne's formula we can see that the locus of point P is a disk with center O and radius $\sqrt{2}R$.

2. *Hint*. Using $a' = PA \sin A$, etc. and substitute.

3. *Omitted*.

Chapter 10 Shum's minimal circle problem

1. As the Figures 1 and 2 show, of the four convex polygons marked as 1, 2, 3 and 4, every two have a common edge but have no common interior. Therefore, the maximum value of n is equal or greater than 4. On the other hand, suppose there are five convex polygons satisfying the conditions in the plane, and one of them is marked as M_0. The other four are respectively marked as M_1, M_2, M_3 and M_4, according to the location of their common point on M_0. As is already known, M_1 and M_3 have a common edge. Therefore, the three convex polygons M_0, M_1 and M_3 either encircle M_2 and leave out M_4 or encircle M_4 leaving out M_2. In any case, it is impossible for M_2 and M_4 to have a common edge. This result is a contradiction. So we infer that $n \leqslant 4$. That is, the maximal value of n is 4.

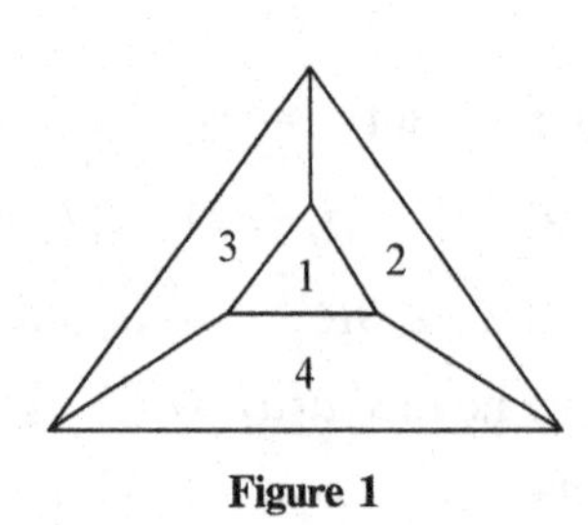

Figure 1

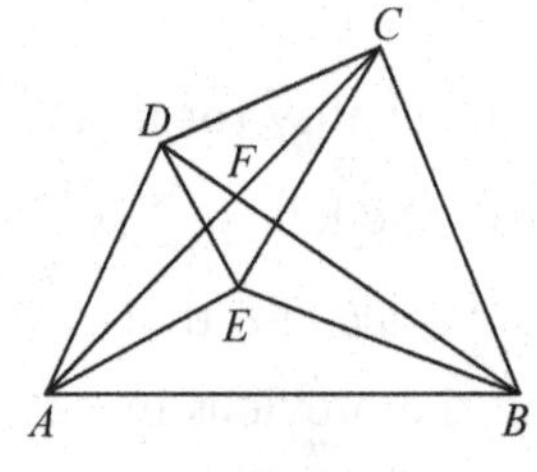

Figure 2

2. The answer is affirmative. We proof it by contradiction. Obviously, there is no such case where three points are co-line. When the convex hull is a triangle, there are two points within it. In this case, the ratio of the maximum distance to the minimal distance between the five points $\lambda \geqslant 2 > 2\sin 70°$, which contradicts what is already known. If the convex hull is a quadrilateral $ABCD$, it is known that E is within it. We may suppose that AC and BD intersect at point F, and E is within $\triangle AFB$, as the Figure 2 shows. And it is already known that $\lambda < 2\sin 70°$.

Here, first let us give two Lemmas.

Lemma 1. For any $\triangle ABC$, we have $\dfrac{BC}{\min(BA, CA)} \geqslant 2\sin\dfrac{A}{2}$.

Lemma 2. If D is a point within $\triangle ABC$, we have

$$\frac{BC}{\min(AD, BD, CD)} \geqslant \{2\sin A,\ \angle A \leqslant 90° \text{ or } 2,\ \angle A > 90°\}.$$

(The proofs of above two Lemmas is omitted.)

From Lemma 1 and our assumption, we see that $\angle AEC < 140°$, and $\angle AEB < 140°$. Therefore, $\angle BEC > 80°$, and $\angle AED > 80° \cdots (1)$. Then from Lemma 2 and the assumption, we see that $\angle ABC < 70°$ and $\angle BAD < 70°$. Therefore $\max(\angle ADC, \angle BCD) > 110°$. Suppose that $\angle BCD > 110°$. Then from (1) we get

$$BC \geqslant \frac{BC}{\min(CE, BE)} \geqslant 2\sin\frac{\angle CEB}{2} \geqslant 2\sin 40°,$$

that is,

$$\begin{aligned}\lambda \geqslant BD &> \sqrt{DC^2 + BC^2 - 2BC \cdot DC \cdot \cos 110°}\\ &= \sqrt{DC^2 + BC^2 + 2BC \cdot DC \cdot \cos 70°}\\ &\geqslant \sqrt{1 + 4\sin^2 40° + 4\sin 40° \cos 70°} = 2\sin 70°.\end{aligned}$$

However, the result contradicts what is already known. So we conclude that the five points are vertexes of a convex pentagon.

3. *Omitted*.

Chapter 11 Inequalities for tetrahedron

1. *Hints*. Suppose that the plane of $\angle PAQ$ intersects all faces of the regular tetrahedron at lines AE, AF and EF. Now we need only to prove that $\angle EAF \leqslant 60°$. And this can be confirmed if it is demonstrated that EF is the shortest side in $\triangle AEF$.

2. *Omitted*.

3. *Hint*. Firstly, consider similar problems on plane and look for proper solution. We need to create a plane π, ensuring it passes through O, the center of sphere, and is perpendicular to the angle bisector OC of $\angle AOB$. Then make use of the symmetry of points, try to prove that points in the plane cannot be on $\overset{\frown}{AB}$. That is, $\overset{\frown}{AB}$ does not go through plane π.

4. Draw line IM perpendicular to the plane $A_2A_3A_4$ with M as the foot, and draw line IN perpendicular to A_3A_4 with N as the foot. If the angle formed by the planes opposite to A_i and A_j is θ_{ij} $(1 \leqslant i < j \leqslant 4)$, we have $\angle MNI = \frac{\theta_{12}}{2}$. Besides, in the right-angled triangle $\triangle IMN$, we have $IN = \frac{r}{\sin\frac{\theta_{12}}{2}}$. Then

$$I_{34} = \frac{1}{2} \cdot A_3A_4 \cdot \frac{r}{\sin\frac{\theta_{12}}{2}}. \tag{1}$$

By the well-known volume formula of tetrahedron

$$V = \frac{2}{3} S_1 S_2 \cdot \frac{\sin\theta_{12}}{A_3A_4}. \tag{2}$$

From (1) and (2), we eliminate A_3A_4 and get

$$I_{34} \cdot V = \frac{2}{3} S_1 S_2 \cos\frac{\theta_{12}}{2} \cdot r.$$

Here notice that $V = \frac{1}{3}\left(\sum_{i=1}^{4} S_i\right)r$. Therefore,

$$I_{34} = \frac{2S_1S_2}{\sum_{i=1}^{4} S_i} \cos\frac{\theta_{12}}{2}. \tag{3}$$

Sum up both sides of the above equation and make use of Cauchy's inequality, we get:

$$\sum_{1\leqslant i<j\leqslant 4} I_{ij} = \frac{2}{\sum_{i=1}^{4} S_i} \cdot \sum_{1\leqslant i<j\leqslant 4} \left(\sqrt{S_i S_j} \cdot \sqrt{S_i S_j} \cdot \cos\frac{\theta_{ij}}{2}\right)$$

$$\leqslant \frac{2}{\sum_{i=1}^{4} S_i} \Big(\sum_{1\leqslant i<j\leqslant 4} S_i S_j\Big)^{1/2} \Big(\sum_{1\leqslant i<j\leqslant 4} S_i S_j \cos^2\frac{\theta_{ij}}{2}\Big)^{1/2}. \quad (4)$$

Then according to the projection formula of tetrahedron,

$$S_1 = S_2 \cos\theta_{12} + S_3 \cos\theta_{13} + S_4 \cos\theta_{14}.$$

That is,

$$\frac{1}{2}\sum_{i=1}^{4} S_i = S_2\cos^2\frac{\theta_{12}}{2} + S_3\cos^2\frac{\theta_{13}}{2} + S_4\cos^2\frac{\theta_{14}}{2}. \quad (5)$$

Multiply both sides of the above equation by S_1 and then sum up, we have

$$\frac{1}{4}\Big(\sum_{i=1}^{4} S_i\Big)^2 = \sum_{1\leqslant i<j\leqslant 4} S_i S_j \cdot \cos^2\frac{\theta_{ij}}{2}. \quad (6)$$

Then by the symmetric mean inequality:

$$\frac{1}{4}\Big(\sum_{i=1}^{4} S_i\Big) \geqslant \Big(\frac{1}{6}\sum_{1\leqslant i<j\leqslant 4} S_i S_j\Big)^{1/2}. \quad (7)$$

Now we can apply (6) and (7) to (4), and finally get the inequality in question.

5. Suppose that for the tetrahedron $A_1A_2A_3A_4$ the area of the face $A_2A_3A_4$ is F_1, and so on. P is any given inner point within the tetrahedron. Then from Cauchy's inequality we have:

$$\Big(\sum F_i\Big)\Big(\sum F_i \cdot PA_i^2\Big) \geqslant \Big(\sum F_i \cdot PA_i\Big)^2. \quad (1)$$

Now vectors from O to the points P, A_i and I are denoted respectively by $\vec{P}$, $\vec{A}_i$, and $\vec{I}$. So $\vec{I} = \dfrac{\sum F_i \vec{A}_i}{\sum F_i}$. Therefore

$$\sum F_i \cdot PA_i^2 = \sum F_i(\vec{P} - \vec{A}_i)^2 = \sum F_i(R^2 + \vec{P}^2 - 2\vec{P} \cdot \vec{A}_i)$$
$$= F(R^2 + OP^2 - 2\vec{P} \cdot \vec{I}).$$

Here, $F = \sum F_i$. As $2\vec{P} \cdot \vec{I} = \vec{P}^2 + \vec{I}^2 - (\vec{P} - \vec{I})^2$, so $(\sum F_i)(\sum F_i \cdot PA_i^2) = F^2(R^2 + PI^2 - OI^2)$. Therefore, form (1) we get

$$\sum F_i \cdot PA_i \leqslant F(R^2 + PI^2 - OI^2)^{1/2}. \tag{2}$$

Now suppose that h_i and r_i stand for the distances from the points A_i and P to the plane F_i, respectively. Then we have $PA_i \geqslant h_i - r_i$. Therefore,

$$\begin{aligned}\sum F_i \cdot PA_i \geqslant \sum F_i(h_i - r_i) &= \sum F_i h_i - \sum F_i r_i \\ &= 4 \times 3V - 3V = 9V.\end{aligned} \tag{3}$$

Combine (2) and (3) together, and make use of the equation $3V = rF$, we get $R^2 \geqslant 9^2 r^2 + OI^2 - PI^2$. And take $P = I$, then the inequality in question is easy to be proved.

Printed in the United States
By Bookmasters